D1433972

THE CANALS OF
SOUTH WEST ENGLAND

THE CANALS OF THE BRITISH ISLES

EDITED BY CHARLES HADFIELD

1. *British Canals. An illustrated history.* By Charles Hadfield
2. *The Canals of South West England.* By Charles Hadfield
3. *The Canals of South Wales and the Border.* By Charles Hadfield
4. *The Canals of the North of Ireland.* By W. A. McCutcheon
5. *The Canals of the East Midlands (including part of London).* By Charles Hadfield
6. *The Canals of the West Midlands.* By Charles Hadfield
7. *The Canals of the South of Ireland.* By V. T. H. and D. R. Delany
8. *The Canals of Scotland.* By Jean Lindsay
9. *Waterways to Stratford.* By Charles Hadfield and John Norris

in preparation

10. *The Canals of North West England.* By Charles Hadfield and Gordon Biddle
11. *The Canals of South and South East England.* By Charles Hadfield
12. *The Canals of North East England.* By Charles Hadfield and Gordon Biddle

Other Books by Charles Hadfield
Introducing Canals
Canals of the World
Atmospheric Railways

THE CANALS OF
SOUTH WEST ENGLAND

by

Charles Hadfield

WITH PLATES AND MAPS

DAVID & CHARLES
NEWTON ABBOT
1967

7153 4176 6
This book is a considerably extended version of part of *The Canals of Southern England* 1955. A companion volume, *The Canals of South East England*, is in preparation

Printed in Great Britain by
Latimer Trend & Co Ltd Plymouth
for David & Charles (Publishers) Ltd
South Devon House Newton Abbot Devon

This book
is respectfully dedicated
to the memory of
EARL STANHOPE, LORD ROLLE
and
JOHN EDYVEAN
promoters of Inland Navigation
and of
JAMES GREEN
civil engineer

CONTENTS

7

8 CONTENTS

ILLUSTRATIONS

9

TEXT ILLUSTRATIONS AND MAPS

PREFACE

NEW source material has been found, and additional research done, in libraries and on the ground, since *The Canals of Southern England* was published in 1955. The old book has stood the test of time well enough, but now that a new edition is needed, it has seemed sensible to divide into two what was already overlong. *The Canals of South West England* is the first of these new volumes; it will be followed by *The Canals of South and South East England*.

This book has been completely rewritten, to incorporate whatever new I or others have found, and to correct what has proved wrong or inadequate in the old account in the light of more recent knowledge. Because the waterways of the south west, except for those linked to the River Parrett, developed separately and hardly influenced one another, I have been able to provide a single account of each.

Those who enjoy canals, present and past, possess a talisman to bring them friends, not enemies, help and not refusals. Inevitable labour has therefore been lightened by the willing assistance of more friends, personal and pen, than can ever be listed in a book's acknowledgements. It is made lighter still when one has a wife and children who are generous to one's interests, and a publisher who is a distinguished transport author as well as an old friend.

CHARLES HADFIELD

CHAPTER I

The South Western
Waterways

✦✦✦

THE waterways of west Somerset, Devon and Cornwall, 150 miles long at their greatest extent, were of little economic importance compared to those of the midlands and north. But the schemes and projects that occupied men's minds between 1769, when Robert Whitworth surveyed a line of canal from the Exe at Topsham to the Bristol Channel at Uphill, and 1870, when a ship canal to cost £3½ millions was proposed from Bridgwater bay to Exmouth, show many varieties of human optimism. Indeed, there were achievements also, small but solid: such practical and useful local waterways as the Parrett Navigation, the Tavistock Canal, or the Liskeard & Looe Union. Of them all, however, only the Exeter Canal is today used for commercial traffic.

Looking back, we can group the waterways of the south west according to the ideas that produced them.

Those built to carry a known local trade big enough to enable their operation to be profitable have always been the most successful. Of this kind were the three earliest in the south west, the Exeter Canal, to by-pass an obstructed part of the River Exe, so that seaborne coal and goods could reach the city, and its products be carried away; the River Tone navigation from Bridgwater and the Parrett, principally to carry coal to Taunton; and the little Parnall's Canal near St Austell, part of the transport arrangements of a tin mine.

There were others: the Par Canal, linked with the mineral-carrying Treffry tramway; the Liskeard & Looe Union, carrying coal, copper ore and granite; the short-lived Cann Quarry Canal near Plymouth; the Tavistock Canal, serving the local copper mining district; and the Stover and Hackney Canals, down which pottery clay was carried to the River Teign to be shipped at Teign-

13

mouth. The partly-built colliery branch of the Dorset & Somerset would also have been of this kind.

The Tone was the first of a small group of navigations that grew up, linked to each other, and based on the minor port of Bridgwater. The Parrett Navigation ran through Langport to Westport towards Ilminster, with a branch to Thorney; the Bridgwater & Taunton Canal replaced the Tone for most traffic to Taunton, though the river remained in some use, both being joined there to the Grand Western Canal through to Wellington and Tiverton; and a branch from the Bridgwater & Taunton at Creech St Michael ran through tunnels and over inclined planes to Chard. At its fullest extent open only from 1842 to 1867, this was the only waterway network the south west achieved. Near it, but not physically connected, the Glastonbury Canal carried what little required to move between Highbridge and Glastonbury.

Because the sea lies on both sides of the peninsula, because Land's End was always a difficulty and danger to small sailing ships, and coal supplies all came from the Bristol Channel side, the idea persisted that a canal should be built to link the two seas, so providing an alternative passage to Land's End, inland towns with good transport to and from ports on either channel, and the south coast with coal. This ruling idea had four manifestations.

The first, and the only one to take partial shape, was for a diagonal line from the Exe across east Devon and west Somerset to the Tone at Taunton, and so to Bridgwater. This got extended higher up the Bristol Channel to Uphill, then farther still to the Avon below Bristol, and then changed its character, as promoters also envisaged the waterway from London via the Kennet & Avon Canal to Bristol being extended on the same scale via Taunton to Exeter, to give a continuous water-line some 200 miles long. The Grand Western Canal from Tiverton to Lowdwells, later extended through Wellington to Taunton, was a fruit of this idea; so was the Bridgwater & Taunton with which it connected, itself a remnant of an earlier line to Bristol.

The second was for a waterway from north to south across Somerset and east Devon, usually from the mouth of the Parrett to the coast near Seaton. Scheme after scheme was promoted, discussed, estimated, planned and subscribed: one, the grandiose English & Bristol Channels Ship Canal, obtained an Act in 1825 to build a waterway for 200-ton craft at a cost of £1¾ millions. But no digging was ever done, and in the end railways and steamships took away the arguments. The other two were less important: for the Public Devonshire Canal in 1794, a sea-to-sea canal

Anno Decimo & Undecimo

Gulielmi III. Regis.

An Act for Making and Keeping the River *Tone* Navigable from *Bridgwater* to *Taunton* in the County of *Somerset*.

Hereas the Clearing and Effecting of a Paſſage foʒ Barges, Boats and other Ueſſels by the Ri-ver Tone, from the Town of Bridgwater in the Coun-ty of Somerſet, to the Town of Taunton in the ſaid County, will be ve-ry Beneficial to Trade, Advantagious to the Pooʒ, and Convenient foʒ the Convey-ance of Coals and other Goods and Mer-chandize to and from the ſaid Towns and Parts adjacent, and will be a means to pʒe-ſerve the Highways lying between the ſaid Towns, which now by continual Car-riages, Carts, Waggons and Wains, are across Devon from the Exe to Barnstaple; and for canals across Cornwall. The American Robert Fulton surveyed one, John Rennie another, Sir Marc Brunel a third. But none had impetus, and none were begun.

Another idea persisted in Devon and Cornwall, born out of agricultural needs and adventurous engineering. Sea-sand from Bude and elsewhere had long been used as manure (as our ancestors called it) to improve the soils of Cornwall and west Devon. From the engineering knowledge of John Edyvean, Edmund Leach and James Green, the energy of Lord Stanhope, the Call family, a group of Exeter merchants, and Denys Rolle, was born the idea of tub-boat canals, of the kind used industrially in Shropshire, but for carrying sand to where farmers needed it. These were to run, not along fertile river valleys, but high up on barren hillsides, lifted to that height by inclined planes, their winding courses no disadvantage because more farms would be served. The Bude and Torrington Canals were built on this principle, and the partly-built St Columb and Tamar Manure Canals would have followed it.

Unlike the waterways of the industrial midlands and north, most of those in the south west were not opened until the 1820s. Even then, few but the corporation-owned Exeter were prosperous; the Liskeard & Looe Union could be relied upon to pay 5 per cent; the Tavistock and the Parrett paid some dividends, the Bude hardly any, and those negligible; the Grand Western, Bridgwater & Taunton, Chard and Glastonbury none but at their winding-up. The St Columb, Torrington, Par and Cann Quarry were privately owned; the last three probably paid their way, though not much more.

The passing of the Act of 1836 for the Bristol & Exeter Railway foreshadowed the end of the Somerset and east Devon group of waterways. The main line was opened in 1844, the Tiverton branch in 1848, that to Yeovil in 1853. By 1868 the Bridgwater & Taunton, Tone, Grand Western, Chard, and Glastonbury were in railway hands, the first two, and the Lowdwells–Tiverton section of the Grand Western, still open, the rest closed. The Parrett Navigation alone survived independently for another ten years.

Farther west, the Liskeard & Looe company decided in 1858 to build a parallel railway, did so, and became a railway concern possessing a largely disused canal; the Tavistock suffered more and more seriously from competition in the years after the South Devon & Tavistock Railway opened in 1859, and probably closed in the eighties; the Stover Canal was sold to a railway in 1862; the

I. Exeter Canal: (*above*) Near Topsham in 1927. The side-lock to the River Exe leads off the canal to the left, by the lock-house; (*below*) *Guidesman* passes Countess Wear swing-bridge in 1957

II. (*above*) A half-lock at Newbridge on the River Tone in 1938; (*below*) the Conservators of the River Tone inspect the Bridgwater & Taunton Canal in 1922

Torrington closed in 1871 so that rails could be laid over part of its bed. Finally, in 1891 the Bude succumbed, partly to railway erosion of its traffic, more because artificial fertilizers had replaced sea-sand unless the latter were sold at prices too low for it to be transported economically, and all but the short broad waterway near Bude was closed.

The following figures, prepared on the same basis as those used in other volumes in this series, tell the story plainly enough.

*Canals and Navigations Open in the South West by type of Waterway**

Date	*Ship Canal miles*	*Broad Canal miles*	*Narrow Canal miles*	*Tub-boat Canal miles*	*River Nav. miles*	*Total miles*
1760	2¼				11⅝	13⅞
1770	2¼				11⅝	13⅞
1780	2¼			4½	11⅝	18⅜
1790	2¼				11⅝	13⅞
1800	2¼	1⅞			11⅝	15¾
1810	2¼	1⅞		2⅜	14⅜	20⅞
1820	2¼	12⅞		8⅜	14⅜	37⅞
1830	5	29½	4⅞	50¾	15⅛	105¼
1840	5	44⅛	4⅞	59¼	28⅛	141⅜
1850	5	44¾	4⅞	74⅝	28⅛	157⅜
1900	5	30½			15⅛	50⅝

On the ground lies the evidences of what was done: the six inclined planes of the Bude Canal, one of which, at Hobbacott Down, was the highest built in this country to carry boats; the vertical boat-lifts of the Grand Western; the tunnels and inclines of the Chard; other boat-lifts on the Dorset & Somerset; the lovely Beam aqueduct on the Torrington; the tiny Morwelldown tunnel, nearly 1½ miles long, on the Tavistock; the twenty-five locks of the Liskeard & Looe Union; or the 70 ft by 20 ft river lock of the Tamar Manure Navigation, the biggest in the south west except for the Exeter ship canal.

In some ways the Exeter Canal is the most remarkable of all, for, setting aside Roman work, it is the oldest in Britain, with three

* The canals and navigations listed are those in Appendix I. I have left out natural rivers such as the Axe and the Parrett below the limits of the Navigation company, but I have included the full length of the Tone. For classification purposes a ship canal is a canal that admits seagoing ships; a broad canal one with locks at least 12 ft wide; a narrow canal one with locks less than 12 ft wide; and a tub-boat canal one taking small boats carrying a few tons each. See *British Canals*, 3rd ed., 1966, for a fuller description.

centuries of history behind it. Twice rebuilt, the present waterway is owned and managed by Exeter corporation, who built the original canal in the reign of the first Elizabeth. Let us turn to it first in this account of the canals of the south west.

CHAPTER II

The Exeter Canal

◆◆◆

THE River Exe runs by Exeter to Topsham, where it broadens to an estuary that is in places a mile and a half wide. Then through a narrow opening between Exmouth and Dawlish Warren, it reaches the sea.

The truth about the blocking of the river is not known.[1] It seems likely that during the reign of Edward I, Isabella de Fortibus, Countess of Devon, built a weir across it, still called 'Countess Wear', which prevented boats going up to the water-gate at Exeter, and that after an inquisition of 1290 navigation was restarted through a 30-ft opening made so that vessels could pass. Between 1317 and 1327 this passage was, however, blocked by Hugh Courtenay, Earl of Devon, whose successors added other weirs and built a quay at Topsham, where goods for Exeter had to be unloaded and pay duty to the Earls. The citizens of Exeter took legal proceedings against the Courtenays, and verdicts were gained, but 'such was their powre and authoritie and such was the iniquitie of those daies as no justice could take place, nor lawe have his dewe course'.[2]

In 1539, after the attainder of Henry Courtenay the year before, the corporation obtained an Act to enable them to remove the three weirs and many shoals that blocked the river. But they failed to make it navigable, and after 1553 nothing was done until, in 1563, John Trew of Glamorganshire was engaged to make a canal alongside the river, the first to be built in Britain since Roman times. His fee was to be £225 and a percentage of the tolls.

The original proposal had been for a canal along the west side of the river on the line of a mill-leat, but the east side was finally chosen. The line left the Exe a quarter-mile from the city walls, beneath which a quay was built and a crane erected, and avoided the river weirs by taking a course that followed the route of the present canal to Matford Brook just below Countess Wear, a length of 3,110 yd, whence the river was improved to Topsham.

Water was directed into it by the building of Trew's weir across the Exe at Exeter, just below its junction with the canal.

The works were begun in February 1564, and opened in the autumn of 1566. The cost, some £5,000, was met by the corporation, and by voluntary subscriptions. It is said that nearly every church in Exeter gave a portion of its plate, and that by this means 900 oz of silver were obtained.[3]

With a top breadth of 16 ft and a depth of 3 ft, the canal took vessels up to 16 tons, which could load their cargoes directly from craft lying in the estuary. Three pound-locks were built, the first on any British waterway, with vertically-rising sluice or guillotine gates, to pass which boats had to lower their masts. There was also a single pair of gates at the seaward end.

The canal, locally called the haven, was sometimes administered directly by the corporation, but during the fifteenth and sixteenth centuries was often leased. There were difficulties in using it: the awkwardness of the approach up the estuary, the fact that it could only be entered at high tide and was not usable at neaps except for very small craft, and the tendency of the river above Trew's Weir at Exeter to silt up. There was also the opposition of those interested in Topsham quay, who tried to collect dues on goods not landed there but entering the canal, until a compromise was reached in 1580, whereby the corporation agreed to make an annual payment in lieu of the duties that Topsham claimed. Three years later they bought the lease of the quay, and held it until it expired in 1614, but failed to get a renewal. Topsham has certain advantages, and though the bulkier imports used the canal, others like linen and canvas probably went by road, as did the serge cloths that were exported.

After the Civil War the canal was in bad condition, as was the approach from the estuary. It suffered also from an awkward character called George Browning, who had cut a leat from just above Trew's Weir to his fulling mill, and often caused a water shortage until his activities were finally stopped. Land carriage therefore increased. Behind these troubles the rivalry with Topsham continued, especially over whether the place should have its own custom house separate from Exeter's. The corporation decided upon improvements, and in 1675 and 1676 Richard Hurd of Cardiff, for a fee of £100, thoroughly dredged the canal, extended it half a mile downwards towards Topsham to avoid a mile of difficult river, built a much larger entrance with a single pair of gates, often called Trenchard's Sluice, that would take 60-ton craft, with a large transhipment basin just inside it, and also a stone

quay and another weir at Exeter. The city was at this time probably the fourth in England, rising to prosperity at the beginnings
of a great trade in serge, and the improvements meant that though
larger craft could not be used, they moved more quickly and with
fewer delays. The result justified the effort: before 1675 the canal
had been leased for £130 p.a.; in 1691 the figure was £800.

Once again there was a decision to enlarge the canal, and in
1698 an agreement was made with William Bayley to deepen it to
14 ft to take seagoing craft.* Unfortunately he absconded in May
1699 with some of the city's money, leaving the canal impassable.
The charges for land carriage at once leaped up. The Chamber
decided to go on with the work themselves, and the citizens, men
and women, flocked to help. Money was borrowed, and the canal
reopened in 1701. It had been straightened and enlarged, being
now 10 ft deep and 50 ft wide, and could take coasting vessels and
small deep-sea craft up to about 150 tons, though this maximum
tended to decrease as it began to silt up. The lower entrance was
still the same, up a narrow and winding side channel from the
river, usable by large vessels only on spring tides.

The three old locks were removed, and Double Locks built instead, a single lock of large size that served also as a passing place.
There was still only a single pair of gates at the entrance, called
Lower Lock, the sill of which was 4 ft lower than at Double
Locks. Therefore the lower pound became equivalent to a lock
2 miles long with a 4 ft rise; consequently it was necessary to raise
the level of the lower pound by 4 ft above that needed to enter the
canal before a vessel could enter or leave Double Locks. There
was now a pair of flood-gates, called King's Arms sluice, where the
canal entered the Exe by the city.

In the years 1715–24 an average of 310 craft a year used the
canal, many of them lighters bringing coal up from ships unable
to get up, and transhipping at Topsham. There was probably little
export trade, and the main imports were coal (35 per cent), cider
(10 per cent), groceries from London (10 per cent), and southern
European goods (10 per cent).

In the year 1750–1, 479 craft used the canal, and exports more
nearly balanced imports than at any other known time in its history. Coal now accounted for 40 per cent of the imports, followed

* Celia Fiennes saw it at this time. It was Topsham, she says, 'from which they
are attempting to make navigeable to the town which will be of mighty advantage
to have shipps come up close to the town to take in their serges, which now they
are forced to send to Topshum on horses by land . . . they had just agreed with a
man that was to accomplish this work for which they were to give 5 or 6000£., who
had made a beginning on it.' *Journeys of Celia Fiennes*, ed. C. Morris, 1947.

by slates (9 per cent), groceries from London (6 per cent), and timber (5 per cent). Cider and southern European goods had fallen to 3 per cent each. Woollens were now a considerable export, together with wheat, oats and flour. Those woollen goods which were destined for London were loaded at Exeter; the others went down the canal in lighters to be shipped from the estuary. In 1795–6, 448 craft used it, of which 158 were seagoing ships and 290 were lighters. The canal was shallower than it had been, and the old estuarial navigation difficulties persisted, yet the greater part of Exeter's trade probably passed by it, though that trade was itself becoming more localized. Coal, wheat and wool were the most important imports, and woollen goods for London and the East India Company's trade, oats, manganese, and paper the principal exports. Receipts, which had averaged £747 between 1751 and 1760,* averaged £2,335½ between 1791 and 1800. Maintenance had cost £33,000 for the 37 years before 1795.

Receipts went on increasing for the next twenty years,† as did the pressure on the canal and the complaints of its users. In 1818 James Green was asked to report. He did so in October 1820, and was commissioned to dredge and straighten the canal, and to repair Double Locks. This work was finished about the end of 1821. In a further report in 1824 he recommended that the canal should be extended to Turf, 2 miles lower down the estuary, where a proper entrance lock should be built. He considered that vessels drawing 12 ft of water could pass the bar at Exmouth and navigate to this point on all tides.

Work began on 20 April 1825. Soon afterwards it was decided to raise the banks of the canal so that it could take craft drawing 14 ft, and to build a basin at Exeter to avoid the difficulties of deepening the river between the end of the canal and the quay. The old entrance having been blocked up, the new canal works were opened on 14 September 1827, and the basin on 30 September 1830. Lastly, as a result of complaints from Topsham, the side-lock into the river there was finished by September 1832. The cost of these improvements had been £113,355, of which £89900, had been raised by the Corporation on mortgage of the canal and the principal revenues. Turf lock had had to be built on piles driven down through clay and underlying bog to the rock beneath, and had cost about £25,000.

The canal could now in theory take vessels of 14-ft draught and about 400 tons burthen, and in January 1834 a brig of 350 tons

* Average of the six years for which figures are available.
† Average for 1801–10, £2,866; for 1811–20, £3,221.

drawing 13½ ft entered the basin. Receipts increased considerably, though they were the product of high tolls and other charges, and were much criticized by users of the canal. They were as follows in two-year averages.

	£
1830–1	6,253
1832–3	6,866½
1834–5	6,921½
1836–7	7,506½
1838–9	8,261
1840–1	8,473
1842–3	8,550½

In 1843 the canal was let for three years at £9,905 per annum. Coal imports had much increased, and colliers could now enter the canal instead of discharging into lighters. Two ships a week berthed from London, and there was both a coastal and a foreign trade, but the disparity between imports and exports had returned. The woollen export trade had gone, killed by war and the growth of northern industry, and though paper and leather exports had grown, they were on a smaller scale.

Here is a description of the basin before the railway came:

I have seen from 20 to 30 vessels, two or three deep, lying there, and the ground covered with various goods and packages. . . . At this time the import of coals was immense, the surrounding towns and villages being supplied by the merchants. There were two trading companies for merchandise having about six vessels each, the tonnage from 120 to 180, sailing to and from London weekly, weather permitting. I have known goods delivered from London within the week, also some delayed more than a month.[4]

The Exeter Canal remains much as Green left it. The Turf entrance lock is 131 ft by 30 ft 3 in, just above it being a basin where the bigger craft could tranship into lighters, and others wait for a suitable wind before passing out through the lock into the Exe. Higher up the canal, but below the earlier entrance, the side lock, able to pass vessels 88 ft 5 in by 25 ft 1 in, enabled lighters and luggage boats to reach Topsham. Double Locks, 312 ft long and 27 ft wide at the gates, but broad enough for two ships to pass, is the only intermediate lock. At the head of the canal, just before it joins the river through King's Arms sluice to give access to the quay and warehouses, a channel to the left leads into the new basin, 900 ft long, 18 ft deep and widening from 90 to 120 ft. Horses had replaced men for towing, and could now use paths on either side of the waterway.

The Exeter, as a ship canal, was in a similar position to the Gloucester & Berkeley; it was an extension to a port, and so was more likely to be benefited than injured by railways if the rails could be made to spring from the canal basin. At Gloucester this was so, and in 1832 an Act had been obtained to build a railway from the canal basin at Exeter to Crediton, but was not carried out. Again, it was the original intention of the Bristol & Exeter Company to end their line at the canal basin, and on 14 March 1836 the Corporation agreed. For some reason, however, they then changed their minds, and though the Act gave permission for it, the line in fact was ended at Red Cow, now St David's, and the canal was deprived of railway communication. Soon afterwards, in 1849, the surveyor to the corporation suggested two possible ways of giving rail access to the basin, and also proposed that it should be linked with St David's station by 5-ton boats passing on the river, and fitted with wheels to use an inclined plane past Head Weir, an interesting parallel to the boats on the Bude Canal. Finally, after protests by traders at the cost of moving goods by road between the basin and St David's station, double transhipment being necessary, and after much pressing by the corporation, the South Devon Railway agreed to build a branch from their line near St Thomas's to the basin. This was authorized in 1865, and opened for goods traffic on 17 June 1867. The Act required that it should be laid in both broad and standard gauge, the latter so that London & South Western trucks could have access to it from their line at St David's. One track, the down, of the main South Devon line between St David's and the junction of the canal basin branch was therefore also laid in mixed gauge.[5]

The opening of the Bristol & Exeter Railway in 1844 at once affected the trade of the canal, and in 1846 the dues were reduced by one-third. As soon as they saw difficult years ahead, the canal creditors got a receiver appointed to collect the income. Thereafter, until 1883 when agreement was reached between the creditors and the corporation, the canal dues were maintained at a level that became less and less attractive compared to the charges at other ports, though reasonable dredging was done.

In 1840 the steamer *Alert* had used the canal. Then, in the autumn of 1847, there was an attempt to bring the fast steamer *Senator* up to Exeter, but she was a few feet too long for Turf lock. In June 1848 another, the *Malcolm Browne*, started a weekly return service between Exeter basin and London for goods and passengers.[6] But, because the steamer exceeded 5 mph and did some damage to the banks, and in spite of the Navigation Committee's plea that 'Steam Vessels for the London trade . . . offer the only chance of successful competition with the railways', the corporation prohibited steamers from using their own power in the canal, and ordered them to be towed by horses, a rule that lasted until 1880. During these years larger steamers largely replaced sailing ships in the coastal and continental trades, and presented the choice of enlarging the canal to a size perhaps similar to the Gloucester & Berkeley, or of limiting the traffic to small coasting craft.

On 22 January 1848, *Woolmer's Exeter & Plymouth Gazette* reported the largest ship to pass the canal, a barque of 330 registered tons, 100 ft by 25½ ft, was refitting in the basin. But this was against the trend; by 1866-7, of 445 ships coming up the canal, only 34 were over 150 tons, and 160 were under 50 tons. The larger ships unloaded at Exmouth, Teignmouth, or Plymouth, whence their cargoes were carried to Exeter by rail. Receipts were down now to £4,103.*

The Exeter Canal Act of 1883 authorized the corporation to redeem the remaining canal debt at 25 per cent of its nominal value, and thenceforward control reverted to them. The number of vessels using the canal, and the receipts, fell steadily to the end of the century, then stabilized, and in the 1960s fell again as oil traffic decreased. The 1901-10 average of vessels was 275; the figure was 256 for 1926, 290 for 1938, 212 for 1952, 277 for 1960, and 120 for 1963. Receipts had by 1901-10 fallen to an average of £1,624. In 1937 the entrance of the basin was widened from 113 to 125 ft, to help larger craft to make the turn.

The Exeter Canal is still open, and is still owned and managed by the corporation. In 1960, 55,431 tons were handled, mainly timber and oil, but by 1963 the figure had fallen to 25,959, at which it has stabilized.

* Average of 1866-70.

CHAPTER III

The Tone and the Bristol & Taunton

✦

THE small port of Bridgwater stands on the tideway of the Parrett, which runs up to Langport and on towards Crewkerne. At Burrow Bridge, not far above the town, it is joined by the Tone, which leads to Taunton and Wellington.

The first attempt to make part of the Tone navigable was made by John Mallett, who in 1638, by a Commission under the Great Seal,

> did, at his very great Expence, make the said River (in some sort) Navigable from the said Town of Bridgwater to certain Mills called Ham Mills in the said County, in Consideration whereof, his late Majesty King Charles the Second . . . did grant to the heirs of the said John Mallett the sole Navigation of the said River, from the said Town of Bridgwater to Ham Mills.[1]

The navigation was in this state when Celia Fiennes saw it, and wrote of the Tone:

> this river comes from Bridgwater even within 3 mile of Taunton, its flowed by the tyde which brings up the barges with coale to this place* . . . and here at this little place where the boates unlade the coale the packhorses comes, and takes it in sacks and so carryes it to the places all about; this is the Sea coale brought from Bristole, the horses carry 2 bushell at a tyme which at the place costs 18d. and when brought to Taunton costs 2 shillings; the roads were full of these carryers going and returning.[2]

It is probable that lack of money hampered John Mallett and his heirs, for in the last decade of the seventeenth century a body of subscribers bought out the Mallett interest for £350, and applied

---

* Coalharbour, near Ham Mills. It may be that at one time boats had got nearer Taunton: see T. Hugo, *A Ramble by the Tone*, 1862.

for an Act to make the Tone navigable to Taunton. The Act stated that after the Conservators, as they were called, had repaid themselves the capital expended in buying out the Malletts and in improving the river, together with interest at 6 per cent, then the tolls were to be reduced, and these tolls, after the cost of maintenance had been deducted, were to be given by the Conservators for the 'only Use, Benefit and Advantage of the Poor of the said Town of Taunton . . .' by building hospitals. This clause was of much importance later, as we shall see.

A second Act was necessary in 1707, and in support of it the undertakers presented a petition:

Showeth that altho your Petitioners have already been in makeing and keeping ye said River Navigable from ye Towne of Taunton to a Certaine Place on ye said River Called Ham Mills att ye Charge of ye Sume of Three Thousand Eight Hundred Pounds upwards yet unless some further considerable Sume be Expended in Building or Erecting an Half Lock or Turnpike below a place called Knapp Bridge on ye said River, in Cleansing a great Shoale called Broad Shoale and doing other Chargeable and necessary Works the said River will never be compleatly navigable. . . .

The supporting petition from two Taunton parishes said that the work on the river had been done

to ye Great benefitt and advantage of both Parrishes and parts adjacent as appeares by Reduceing the Charge of Land Carridge from Bridgwater to Taunton to neare half the Prise and by keeping Coales to neare as low a value as they were att before a duty was laid thereon. . . .

By 1717 the river had been made navigable to Taunton for barges carrying something over 15 tons, with one lock and at least two half-locks (single pairs of gates penning back the water, the equivalent of the old staunches), and in 1724 an agreement was signed to build a horse towing path from Ham Mills to Taunton. The total length of the navigation from Taunton was 11⅝ miles to Burrow Bridge and another 6 to Bridgwater, but the main work was, as we have seen, needed upon the last 3 miles to Taunton.

In 1724, about forty years after Celia Fiennes, Defoe saw the river, and described it as 'infinitely advantagious' to the people of Taunton.

first by bringing up coals, which are brought from Swanzy in Wales by sea to Bridgwater, and thence by barges up this river to Taunton; also for bringing all heavy goods and merchandizes

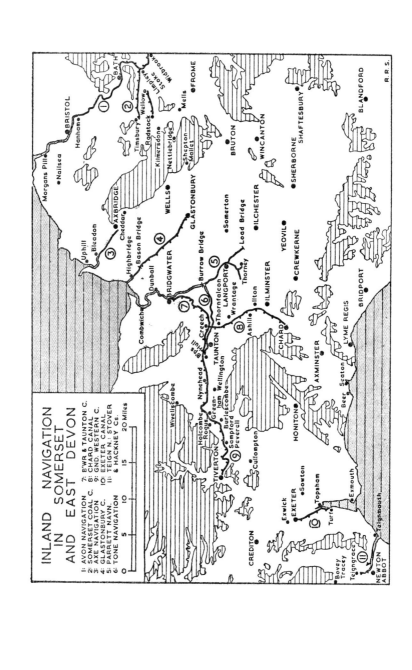

INLAND NAVIGATION
IN
SOMERSET
AND EAST DEVON

1: AVON NAVIGATION   7: B.W.R & TAUNTON C.
2: SOMERSET COAL C.   8: CHARD CANAL
3: AXE NAVIGATION   9: GND. WESTERN C.
4: GLASTONBURY C.   10: EXETER CANAL
5: PARRETT NAVN.   11: TEIGN N.: STOVER
6: TONE NAVIGATION   & HACKNEY Cs.

0   5   10   15   20 Miles

R.R.S.

from Bristol, such as iron, lead, oyl, wine, hemp, flax, pitch, tar, grocery, and dye stuffs, and the like. . . .[3]
In the same year of 1724, Dr Amory, a Taunton poet, wrote as follows of his river:

> *The fattening Tone in slow meanders moves,*
> *Loath to forsake the happy land it loves;*
> *Forced to the main, by Nature's laws, it bears*
> *Back floating vessels fraught with richest wares;*
> *And diff'ring products, from earth's diff'ring shores,*
> *Gather'd by commerce, lavish on us pours.*[4]

The debt for buying the Malletts' rights and building the navigation amounted in 1717 to £5,697 8s 10d, and dividends were subsequently paid upon that sum.* It was considered that the Conservators were entitled to 6 per cent upon this sum under the Act of 1699, and because earnings did not then allow it to be paid, 6 per cent less the year's profit was added each year to the total of the debt. The nominal debt therefore ceased to have any relation to the actual sum spent on the navigation. By the time the river was in fact earning 6 per cent on the original debt, the nominal debt had increased so much that it was not possible to earn 6 per cent on that; therefore the nominal debt kept increasing and, as we shall see, later got quite out of hand. By 1760 it had reached £14,431 13s 8¾d.

The following table shows the tolls and the sums paid in dividends in five-year averages from 1708 to 1763, and illustrates a slow but steady growth of trade.

Mid-June	Tolls			Dividends		
	£	s.	d.	£	s.	d.
1708–9 to 1712–13	321	14	10¼	—		
1713–14 to 1717–18	379	13	6¾	—		
1718–19 to 1722–3	379	3	9¾	263	18	3 (4 yrs)
1723–4 to 1727–8	395	6	4	263	11	1¼
1728–9 to 1732–3	391	10	11¾	253	19	11
1733–4 to 1737–8	413	6	11¼	303	18	3¼
1738–9 to 1742–3	404	2	4	288	11	8
1743–4 to 1747–8	429	5	6	333	1	9
1748–9 to 1752–3	426	14	4	315	14	10½
1753–4 to 1757–8	444	0	10	357	7	3½
1758–8 to 1762–3	464	12	0½	370	7	2

A dividend of 5 per cent needed £284 15s 0d, so dividends

* The word 'dividend' sounds curious in connection with 'debt', but both words were used in the River Tone records.

rose during this period from rather under 5 per cent per annum to rather over 6½ per cent on the original debt. Later canal undertakers would have envied the low cost of maintenance.

In 1735 James Drew of Taunton, goldsmith, applied to the Somerset Commissioners for Charitable Uses for a decree that the undertakers had been repaid their original money in terms of the Act, and that the poor of Taunton should therefore benefit. Such a decree was made, but the Conservators won on an appeal heard in 1738.

It was not till 1751 that the Committee were empowered to appoint a superintendent,

> a proper Person or Persons to keep the Locks in repair and the River navigable for a Term of Years, and empower'd to contract for a yearly Salary to be paid for that purpose.

By the beginning of the canal age the river had been made fully navigable by means of locks and half-locks from Coalharbour upwards.

From 1760 to the end of the century the Conservators of the Tone controlled their waters with little more to trouble them than:

> Whereas a quantity of bad half pence has been tendered for the payment of the Tolls on the River Tone it is Resolved . . . the Clerk . . . do not take or receive of any person or persons for the Tolls at the Lock any greater sum in half pence at any one payment, than one shilling, and that he be carefull to take good half pence.[5]

or

> That none be invited to accompany the Conservators on a View of the River but Proprietors and Tradesmen: and that Expences of the Dinner on the Survey, be limited to an allowance of Six Shillings per head for the Ordinary & Extraordinary: and that any Expences incurred beyond that Proportion be defrayed by the Persons Present.[6]

There was, however, a flood which caused the Conservators to go without dividend in 1764 and 1765, and in the former year it was recorded:

> this year made good ye ground work at Barpool lock in which was used in filling up what was wash'd away under ye walls and Gates and Repairing ye walls upwards of 200 Tunn of Stone. . . .

A note of the following year shows that the locks had timber booms across the bottom, filled in with stone, and sides probably of stone:

> . . . this year at Ham made ye ground Timber all new all across

ye pound with 4 Booms, and fill'd between such boom with Stone Repair'd ye walls and made 4 new gates.

In April 1793 we are given an incident of river life:

Thursday the 25th as one William Barrington, a boatman, returning from Taunton with a boat to Bridgwater, to which former place he has been with coal, about one mile from the coal yard, meeting another boat coming up loaded, a dispute happened which should pass first; the deceased persisted he would pass the locks, and standing on the fore part of the boat, his foot slipped, his head got between the hatches, which immediately fell fast, his brains were dashed out, and he died instantly.[7]

Takings and dividends rose rapidly as agricultural prosperity showed itself in the growth of demand in and around Taunton for coal: by the end of the century the river was carrying about 14,000 tons of goods a year, of which about 11,500 tons was coal upwards from Bridgwater.

	Average Receipts £
1763–4 to 1767–8	520
1768–9 to 1772–3	584
1773–4 to 1777–8	676
1778–9 to 1782–3	650
1783–4 to 1787–8	694
1788–9 to 1792–3	794
1793–4 to 1797–8	982
1798–9 to 1802–3	1,137

And in 1797 the dividend was 2s 3½d in the £, or about 11¼ per cent. In spite of this, the nominal debt continued to rise astronomically because of the way in which it was calculated, till in June 1800 it had reached £85,466, it having increased by £4,876 since the previous year. In 1800 the Justices, to whom the accounts had to be submitted, refused to examine them, presumably because it had been represented to them that because of the Conservators' book-keeping the poor of Taunton were each year getting further away from any benefit from the navigation, while the Conservators were getting larger and larger dividends. The Justices demanded that the amount of the debt should be recalculated at a reasonable figure, and should then be paid off, so that tolls could be reduced and the poor eventually benefited. The Conservators, after making some threats to expose irregularities in the handling of the magistrates' own expenses, made a series of offers of revised debt totals, each lower than the last, until final agreement upon £13,000 was reached in 1803, which was confirmed by Act of Parliament. Interest of 6 per cent only was thenceforward to be paid, and the balance of income applied to reduce the debt, which by 1827–8 stood at only £4,426.

At this time the Tone seems to have had four full locks, at Obridge, Bathpool, Creech St Michael and Ham, and four half-locks (i.e. a single pair of gates), one at Bathpool, below the full lock, one at Ham about 450 yards below the lock, and two at Newbridge 100 yards apart, the lower of which was called Curry Moor.

In December 1792, a meeting was held at Wells with John Billingsley in the chair to support a Bristol & Western Canal from the Avon at Morgan's Pill near Bristol to Taunton. It decided to confer with those who were at that time promoting the Grand Western Canal from Taunton to the River Exe at Topsham, and also, frankly recognizing that hands helping them towards a future Act must be rewarded, ordered that

> one-fourth part of the number of shares be reserved to accommodate the Members of the county and High Sheriff thereof, and for landowners through whose estates the said canal shall pass, and such other gentlemen whose assistance by the Committee be deemed necessary to forward this undertaking.[8]

Quick profits could of course be made in a speculative period by subscribing for scarce shares and then selling them, deposit paid, at a profit.

A month later another meeting at Taunton revived the Taunton–Uphill section of the old project that Whitworth had surveyed, as another means of extending the Grand Western to the Bristol Channel. The project was partly the result of annoyance with the proceedings at Wells, which were described as 'highly improper and exceptionable, as well because sufficient notice . . . was not given, as because the object of it was so studiously concealed'.*[9]

Both projects went forward. The Bristol & Western was surveyed by John Longbotham and William White, who chose a 40-mile line running near Uphill, with only one lock, almost coincident with that of the Taunton group. The former then suggested a union of the two sets of promoters, but when discussions took place, the latter were so annoyed at the minor part they thought themselves to have been offered that they withdrew, and kept to their limited Taunton–Uphill plan. Both now tried to get support from the Grand Western. The Taunton & Uphill, meeting on 7 November 1793, resolved: 'that to concur and cooperate

---

* See my *British Canals*, 3rd ed., 1966, p. 109, for a further account of this incident.

III. Bridgwater & Taunton Canal: (*above*) Passing through Bridgwater; (*below*) Bridgwater dock. The canal entrance is in the foreground, that to the Parrett in the distance

IV. Chard Canal: (*above*) At Wrantage, looking towards Crimson Hill tunnel, 1961; (*below*) the north portal of the tunnel, 1967

with the designs of the Grand Western Canal Committee will be to both parties a scheme of general utility',[10] whereas six weeks later the Bristol & Western 'resolved that this Committee do consult and mutually cooperate with the Grand Western Canal Committee, agreeable to their request'.[11] It seems that the Grand Western favoured the Taunton & Uphill.

In the autumn of 1793 the Bristol & Western got hold of William Jessop to check their line; by the end of the year he had approved it with certain changes. Oddly, Jessop then moved on to the Taunton & Uphill, where he checked a survey previously made by Josiah Easton. That group then sent a delegation to Newport to investigate coal prices, and thought of extending beyond Uphill to the Nailsea collieries, and of negotiating with the Conservators of the Tone.

The Bristol & Western, now getting cold feet, decided in June 1794 to drop the Nailsea–Bristol portion of the line from their bill 'till other more favourable circumstances than appear at present occur'.[12] But then they received encouragement from the Chard Canal inter-Channels project (see p. 38) with its promise of additional traffic, negotiated a junction between the two lines, and again thought the whole length practicable.

By now, however, landowners, who were faced with three canal projects in north Somerset which might affect the drainage or water supply of their lands, started to organize opposition. They may have helped to finish off the Taunton & Uphill, who deposited a plan in 1794, before disappearing into the canal promoters' twilight. The Bristol & Western kept on as far as a bill in 1796 for their line from Morgan's Pill to Taunton. It was defeated and dropped, the promoters only meeting in November to settle their debts.

These projects for linking Taunton with the Bristol Channel had come to nothing, but that from Taunton to the Exe had been authorized in 1796 as the barge-sized Grand Western, to join the Tone a little above the town bridge, and to incorporate 500 yards of the river in its line. In April 1810 work on it began near Holcombe Rogus, with John Rennie as engineer. The Kennet & Avon from the Thames at Reading to Bath, whence it fell by a flight of locks into the Avon that led to Bristol, Rennie's also, was by now nearly finished, and would be opened in December. If the Avon navigation could be avoided by a canal extension from the top of Bath locks to Bristol, and if another canal were to be promoted thence to the Tone at Taunton to join the authorized line of the Grand Western, London and Exeter would be con-

C

nected by inland navigation: though whether anyone would have used it throughout against the competition of cheap coasting craft is more than doubtful.

Rennie automatically became engineering consultant to both projects. During 1810 he surveyed and reported on the Bath & Bristol Canal scheme, and also on the proposed Bristol & Western Union, soon to be called the Bristol & Taunton. Of the latter he wrote:

> as the Canal is intended to form a Junction between the Grand Western Canal now making and the City of Bristol by the intended Bath & Bristol Canal with the Kennet & Avon Canal at Bath which is nearly compleated, it ought to be made of the same size i.e. for Barges of fifty Tons burthen.'[13]

But he was at the time involved in one more scheme, that for a canal across Somerset for 120-ton coasting vessels from the Bristol Channel at the mouth of the Parrett to Seaton (see p. 39), which would be likely to cross the Bristol & Taunton near Bridgwater. Rennie had an exciting canal empire in prospect.

Those behind the Bristol & Taunton included Sir George Yonge and the Heygates, supporters of the Grand Western, and John Ward of Marlborough, of the Kennet & Avon's firm of solicitors. The promoters' own lawyer was Isaac Cooke of Bristol, whose firm was to be long connected with Somerset canal building. Rennie told them that 'no Line of country can be more favourable for a Navigable Canal'. He proposed one on the level from a lock at Morgan's Pill on the Avon near Portbury below Bristol, to run north of Nailsea, through a 600 yd tunnel at Clevedon and a longer one of 1,050 yd at Banwell to the south side of the Parrett at Bridgwater, where a short branch would link it with the river, 'unless the Canal from the English Channel to the Severn shall take place, or unless the Inhabitants of Bridgwater shall make a Wet Dock there,' when a junction should be made. The canal would then rise by two locks to cross the Parrett near Huntworth, and run by Creech St Michael to join the authorized line of the Grand Western in the Tone at Taunton. The Axe and Brue, like the Parrett, were to be crossed by aqueducts, and there was to be a 2-mile branch to Nailsea and one of six to Cheddar.

There were, of course, objections. On 21 January 1811 a meeting of landowners was held at Bridgwater, and also a Session of Sewers of much the same membership, to oppose it because it would injure landowners and occupiers and hazard drainage, while the existing facilities for carriage between Bristol and Taunton were 'in every respect preferable to the Proposed Canal, which did not pass through country where manufacture and commerce were carried on'. The Tone Conservators also objected, on the grounds that their trade would be seriously affected.

However, like the Bath & Bristol, the Bristol & Taunton obtained an Act in 1811, though without the Cheddar branch. Rennie had estimated it at £410,896; in the enthusiasm of 1810 £571,800 had been offered; but the Act[14] authorized a capital of £420,000, with £150,000 more if necessary. Those concerned with central Somerset drainage had inserted an onerous provision that the company could not cut any part of the section between Clevedon and the Parrett until they had built the portions on either side, and yet must complete the central section within four years. As for the Tone, the company had within three months of the Act to agree with those who owned shares in the Tone debt to buy them, and also to purchase the rights owned by the overseers of the two Taunton parishes mentioned in the Act of 1699. While the bill was going through, indeed, these two vestries agreed to accept £10 pa each in settlement of their claims.

A few moves were made, and then the company lay quiet for a time. About 1818, however, they bought from the Grand Western, who were not likely to reach Taunton in the foreseeable future, some shares in the Tone debt that company had acquired, and went on to buy others until by 1822 they held the whole debt. This did not, however, enable them to control the appointment of Conservators, who co-opted each other.

After several efforts, a shareholders' meeting was got together in March 1822, stimulated by contemporary activity in inter-channel schemes. By this time, of course, the powers to build the central section had lapsed, and with them any point in going on with that from Morgan's Pill to Clevedon and Nailsea. But that from the Parrett at Huntworth to Taunton offered an acceptable alternative to the awkward Tone navigation, could be viable by itself, and might give a link to the proposed inter-Channel canal. Or so the meeting must have thought, for they authorized a call on shares for this purpose. By this time, of course, many of the original subscribers had died or were untraceable, and calls could only be made on 712 shares out of the original 4,200.

By the end of July tenders were out, and in September, after another call, the newspapers carried an encouraging item:

> Upwards of 200 labourers, called navigators, lately employed on the Bude Canal, have arrived in Taunton for the purpose of commencing operations on the projected Taunton and Bristol

Canal. They are immediately to open the ground at Fire Pool Weir, near that town. A party of the workmen are also to commence their labours at Creech, and a bridge, near the present one at Bathpool, is to be erected. The line will terminate in the Parret, about a mile from Bridgwater.[15]

This impressive contingent only laboured for a few months, because Mr and Mrs Gray of Buckland Farm, Durston, landowners on the canal route, obtained an injunction against the company on the grounds that they had not fulfilled the conditions of the 1811 Act, and therefore their powers had expired. After an attempt to dissolve the injunction had failed early in February, the company at once decided to seek fresh powers. The bill was ready early in 1824, and in March a Conservators' delegation went to London to seek a Bill of Injunction and to petition against the scheme. In a published document they committed themselves to the statement that the Tone was 'a good Navigation, not only adequate to all the present wants of the Traders, but to any possible extension of them'.[16] On 18 May they heard that the bill was through the Commons, and 'it appearing to this Meeting that the Bill if passed into a Law will destroy the River Tone Navigation', decided to oppose it in the Lords. Their opposition was later withdrawn on an understanding that they would be compensated for any prejudice to the river, and the bill[17] passed during 1824.

It authorized variations in the Parrett–Taunton section of the old line, and enabled the company to lock down into the river at Huntworth and build a basin there. The original capital powers were retained, but the rest of the line was abandoned, the company's name being changed to the Bridgwater & Taunton. The new company then asked the Grand Western to extend their line to Taunton, or to allow them to build the intervening section themselves, but before anything could be done, the Ship Canal company took a hand.

CHAPTER IV

English and Bristol Channel Schemes

The idea of linking the English and Bristol Channels by a canal runs through the seventy years from 1769. Many plans were advanced, much preliminary work was done and money spent, but only one was begun. Eventually the success of railway transport and the application of steam to the coasting trade made it less desirable to create an alternative passage to that round Land's End, source of many delays and some disasters in sailing days. Sometimes a ship canal to carry small coastal vessels was intended; sometimes a barge canal to which goods brought by sea would have to be transhipped; once, indeed, a tub-boat canal. Some were thought of mainly as connecting the channels and only incidentally as supplying the country through which they were to run; some the other way about. But all had in common a sea to sea route. In this chapter I have described the schemes in Somerset and east Devon; the partly-built Grand Western has a chapter to itself; there were others in Devon and Cornwall which we shall come to later in this book.

In 1768 a group of Taunton men asked James Brindley to survey a line for an inter-Channel canal. The work was done in 1769 by Robert Whitworth under his supervision, the lines chosen being from Topsham on the Exe through Cullompton, or from Exeter by Cullompton or Tiverton to Wellington and Taunton. Thence barges would use the Tone Navigation to Burrow Bridge, where a second canal would run by Bridgwater, Glastonbury, Wells and Axbridge to Uphill near Weston-super-Mare. Nothing then came of this Exeter & Uphill Canal.

At the same time as he had surveyed the Exeter & Uphill line in 1769, Whitworth had studied another route across Somerset from the Parrett near Langport to Seaton, which did not, however, enter the sea. During the Canal mania of the early 1790s,

Whitworth himself was called back to re-survey this old line to the Parrett, which he again found practicable. In August 1793, the idea was revived as the Chard Canal, not to be confused with the Chard Canal authorized in 1834 and later built.

Another survey was then made by Josiah Easton, and a rather different and more extensive route was worked out. This was proposed as a line from the English Channel at Axmouth near Seaton, via Axminster, Chard, Ilminster, Creech St Michael, Bridgwater, Huntspill and Congresbury to the collieries at Backwell near Nailsea, where it would join another from Backwell via Yatton to Uphill, which has also been suggested, perhaps by a different set of promoters. There would be a branch from the main line near Chard to Crewkerne,[1] and another from West Hatch to Ruggin near West Buckland, Wellington.

These plans were brought before a meeting on 29 August 1794, and approved; a committee was appointed, and a decision taken to apply for an Act. Because the proposed line covered part of the same course as the Bristol–Taunton project, many meetings and much bargaining ensued between the promoters of the Chard, the Bristol & Western, now called the Bristol & Taunton, and the Grand Western. A year later the Chard agreed to join the Bristol & Taunton near Taunton,[2] still with the Crewkerne and Ruggin branches, the canals to be of the same depth of 5 ft and width of 30 ft at top The Chard promoters evidently thought that this agreement for a through English–Bristol Channels line might eliminate the Grand Western scheme, for they reported in August 1795 that their canal 'might be further extended to Wellington, and to or near the town of Tiverton . . . and from thence to other places'.[3] However, it was the other way about. The Grand Western committee obtained an Act, and the other two plans lay dormant.

This Chard Canal plan, renamed the English & Bristol Channels Canal, was revived at a meeting at Chard in October 1809, which followed a dinner given by Mr T. Pyke, who had been concerned with the earlier project, and thought the time ripe to revive it, and the issue of a particularly moving prospectus, which after explaining the great commercial benefits, high profits, and low cost of such a canal, ended:

> A Family whose firing now costs Twenty Pounds per Annum, will procure the same for Ten, and the Poor, (who from the great and increasing scarcity of wood and other fuel, throughout the interior of the country, are often driven to the parish officer, or to robbery for this necessary article of human

comfort, to warm in the severity of winter their shivering limbs;) will obtain it in plenty and at a reasonable price.

At the meeting a canal from the Parrett by Chard to Seaton was suggested, to cost £70,000 and to bring in £30,250 a year. There was later some consideration of a railroad, after which in January 1810 Rennie was asked to survey for a canal. The work was done by W. Bond of Axminster and J. Dean of London under his direction (though in 1824 Bond and Dean were to call themselves the original projectors), and a report written in July 1811. Rennie rejected the idea of a barge canal, because of the necessary transhipment of goods passing from one channel to the other, and recommended a small ship canal for vessels of 120 tons, a figure which would cover most of the coasting craft of the time. This should start, not in the Parrett, but from a wet-dock at Combwich at its mouth, and run by Bridgwater and Langport, up the Parrett vale to a summit near Chard, and down the Axe valley past Axminster to Seaton, where there would be another dock, with a resort harbour at Beer. Except the northern section, his line south of Langport closely followed Whitworth's of 1769. The estimate was £1,330,084, of which £150,000 was for the resort harbour at Beer.

The surveyors, Bond and Dean, later wrote:

the surveys, plans, sections, etc. . . . were submitted to the consideration of a Committee of scientific men, and approved by several Peers, Members of the House of Commons, Landowners, Ship-owners, Merchants, Coal, Copper and Ironmasters, and others of the first respectability, at meetings held in London, Bridgwater, and Chard, subscribers to the amount of £900,000 and upwards, but who, on account of the then scarcity of money, and the general pressure of the times, agreed to postpone all further proceeding in the matter until a more favourable period should arrive.[4]

The progress of the Grand Western probably also had something to do with the decision to keep a permanent committee in being, but not to go forward, and to share out the funds already raised.

The next move was not made until 1821, when Mr Pyke circulated copies of Rennie's report of 1810. This seems to have led a different group to go back to the older plan for a barge canal much more for local use than for inter-Channel trade, and to employ James Green to make the survey. Green was at this time just completing the Bude Canal, mostly for small tub-boats and using inclined planes instead of locks. By this time the Bristol & Taunton project was, as we have seen, active again at the Bridgwater and Taunton end, and Green therefore suggested a tub-boat canal to run from the Tone, or the Bristol & Taunton Canal 2 miles from Taunton, to Beer. There were to be five inclined planes, four tunnels, and a pier at Beer, to which the boats would descend by an extension of the rails of the last plane, and whence their contents could be tipped into waiting ships by upending the boats. He concluded his report:

> I am of opinion that a Canal on a very small scale, on which boats of four or five tons may be navigated in sets of four, six or eight boats, drawn by one Horse on the level Ponds of Canal, and passed singly up and down inclined planes by the aid of water Machinery, instead of locks, will best answer the intended purpose.

The cost was put at £123,156, the yield at 24 per cent. Notice of intention to seek a bill was published in September 1822. A meeting at the Angel Inn at Chard on 18 December accepted his report, subscribed a good deal of money, decided to apply for a bill, and appointed a committee, which included Lord Rolle and John Thomas of the Grand Western and the Bristol & Taunton. Amidst brisk controversy about the proposed English & Bristol Channels Junction Canal, as it was called, effort was switched from Somerset to London, whence it was hoped much of the money would come. Pares & Heygate, London bankers, but having the Heygates's interest in the Grand Western, were given the projectors' account; London solicitors were joined to Mr Thomas E. Clarke of Chard, and the committee now gathered there. A meeting in London in February, with Sir Thomas Lethbridge in the chair, supported Green's plan, and soon afterwards a bill was introduced.

By April it had been withdrawn, for times were good, canal and railway speculations were rising throughout the country, and opinion was swinging towards a ship canal, though Green's tub-boat plan remained in memory and was the clear begetter of the Chard Canal of the following decade. Mr Pyke, and also Bond and Dean, was still supporting the Rennie plan, and in 1823 said of Green's proposal, 'a more rediculous (sic) ill-advised, inefficient, incompetent scheme was never devised', and Pyke followed this broadside with his own statement that a scheme on Rennie's lines might be carried out for about £600,000, 'but if it were five times as much, its wonderful productive tolls would be amply sufficient to pay every expence of its execution, and leave fifty per cent interest to the shareholders'.[5]

This statement may have helped the optimistic movement that culminated in a meeting at the London Tavern, Bishopsgate Street, on 9 June 1824:

to take into consideration a Plan for making a Ship Canal between the Bristol and the British Channels, in order to avoid the dangerous navigation round the Land's End, especially in winter. Sir T. Lethbridge took the chair of the meeting, and spoke strongly in favour of the plan. Mr Pollock said that, in going round the Land's End, in the last three years, there had been a loss of property to about £300,000. Sir T. Lethbridge said, the loss of lives was estimated at 200 per annum. A string of resolutions, expressive of the advantages of such a canal, and others for entering on the undertaking, were all agreed to unanimously, and the meeting dispersed though under the impression that the cost would not exceed a million pounds.[6]

This time Thomas Telford was commissioned to make the survey, helped by Captain Nicholls, later to become Sir George, chairman of the Gloucester & Berkeley Canal Company and then of the Birmingham Canal Navigations, and James Green, who actually signed the plans. They used the levels of Green's earlier scheme, but even then they must have moved quickly, for their preliminary report was published in mid-August, and a subscription list immediately opened. The full report followed in December, and was adopted at a meeting at the London coffee-house on the 16th. The engineers chose a line to the west of the 1809 route, and proposed a ship canal 15 ft deep, with 30 locks, to take vessels of 200 tons, on a route from Stolford to Beer via Creech St Michael, Ilminster and Chard. The cost, including two harbours, was estimated at £1,712,844; the revenue at £210,847 pa, and the expenses at £22,000 pa.

Opposition broke out. A meeting of landowners and others at Axminster in January 1825, among other resolutions passed the following:

That it appears to be one of the most wild, and visionary schemes ever laid before the Public, and calculated only to benefit those who will participate in the money of the subscribers.

That, without intending any imputation on Mr. Telford and Captain Nicholls, it appears to this Meeting, that those Gentlemen, from being unacquainted with the peculiar soil of the country, have proceeded on erroneous grounds . . . and that their Reports in other respects . . . are founded on the most fallacious and exaggerated statements; and that, so far from the proposed canal being of any public or private benefit, it is im-

SHIP CANAL,

FOR THE

JUNCTION

OF THE

ENGLISH AND *BRISTOL CHANNELS.*

REPORTS

OF

Mr. TELFORD AND Captain NICHOLLS;

WITH

PLANS ANNEXED.

Published by Order, at the General Meeting of Subscribers.

LONDON:

PRINTED BY S. BROOKE, PATERNOSTER-ROW.

1824.

3. Telford & Nicholls's report of 1824

possible that it can ever produce the least return whatever to the Subscribers.

The meeting, after comprehensively showing the injury likely to the public, the farmer and the landowner, ended by producing a guaranteed winner; that the canal

in a National point of view, will be still more injurious, by destroying that great nursery for hardy and intrepid Seamen, which, it is well-known, the navigation of our Southern Peninsula has been chiefly instrumental in forming.[7]

They decided to oppose the Ship Canal bill, and a pamphlet was issued which calculated the true receipts at £31,996, and expenses at £42,000. It also more than doubled Telford's estimate for land.

On the other side, Alexander Baring, M.P. for Taunton, may be taken as an example of a supporter. He wrote on 18 April 1825:

The more I consider the general question the more strongly I am convinced of two opinions I all along subscribed. That the Canal will be of immense public benefit, & it will cost at least twice the sum it is estimated. I attended the Meeting & I have since promised to subscribe 50 Shares, not as you may suppose from an expectation of improving my fortune but from thinking an adventure of such great importance wanted a little help.[8]

The Bridgwater & Taunton's committee now saw the chance of a take-over. After brisk negotiations, they reached agreement on 28 March 1825 upon the substance of a clause to be inserted in the Ship Canal bill. The latter company consented to buy the Bridgwater & Taunton on 31 March for £90,000, plus £7,307 spent on buying shares in the Tone debt. This sum was to be given for the unfinished canal as it then was, together with the company's land and construction equipment. Payment was to be made in three instalments within nine months of the passing of the Ship Canal's Act. Meanwhile the Bridgwater & Taunton committee were to continue building their canal on a repayment basis for expenditure not already covered in the purchase price. Baring wrote to the Conservators: 'You no doubt know the agreement made with your old enemies the Bristol Company who made a good escape out of a wretched adventure.' The Conservators' own opposition to the bill had taken the form of insisting that the Ship Canal Company should build a full-sized branch to the town, an idea the townspeople naturally supported. This was conceded, presumably for inclusion in an amending Act, so that Taunton now looked like having three waterways from the Bristol Channel to itself.

The Grand Western, however, unsuccessfully opposed the proposed purchase of the Bridgwater & Taunton by the Ship Canal

OBSERVATIONS

ON THE

PROJECTED SHIP CANAL,

FROM

STOLFORD,

IN THE COUNTY OF SOMERSÉT,

TO

BEER,

IN THE COUNTY OF DEVON,

PUBLISHED BY DIRECTION OF SEVERAL GENTLEMEN OPPOSED
TO THE MEASURE,

SHERBORNE:

PRINTED BY HARKER AND PENNY, MERCURY-OFFICE

Company. They doubted the feasibility, even more the success, of the Ship Canal, and were afraid that a small scheme of great use to them would disappear into a large and financially unstable one. They only succeeded in getting a clause that the Bridgwater & Taunton should still be finished within the statutory three years.

The Act[9] for the English & Bristol Channels Ship Canal was passed, authorizing a capital of £1,750,000, close to Telford's estimate, the Act's powers not to come into force until the full sum had been subscribed, and £750,000 more if necessary. 'Never in the recollection of the oldest inhabitants of this flourishing town was any measure of a local nature anticipated with such intense anxiety and lively feeling,' wrote of a Taunton paper.[10] On Wednesday, 6 July, when the royal assent was expected, dense crowds filled the streets of Chard from midday. At a little after six in the evening, a messenger arrived in a carriage with the news that Mr Salter, managing clerk to Mr Clarke the company's Chard solicitor, whose firm had been concerned with the inter-Channel project since 1793, was following with news of the bill's passing:

> thousands of persons proceeded to meet him, and after taking the horse from the carriage and decorating it with laurels and flags (preceded by the Chard band playing 'see the conquering hero comes', the beadle . . . and several hundred respectable inhabitants of the town) drew him to the Market Place.

Several hogsheads of strong but free beer enabled the locals to drink the health of Lord Rolle, Lord Poulett and all concerned with the scheme.

Subscriptions of over £1½ m had been obtained, but the boom conditions of the time ended, and with them the canal's prospects. A shareholders' meeting was held in August, and in the autumn of 1825 James Green and others were surveying for the Taunton branch, but after that action ceased, until in 1828 the company despairingly reverted to a variation on Rennie's scheme, now to take vessels of 100 tons and to run from the south coast to the Tone at Burrow Bridge, at a cost of £600,000. After that, the company fades out of existence.

But not the idea, which got translated to rails. A Bristol to English Channel Railroad, a horse-drawn line, had been suggested as early as 1809.[11] But in 1833 a *Prospectus for a Rail Road from Minehead on the Bristol Channel, to the English Channel*, dated 30 September, was sent out by an unnamed person from Carhampton near Dunster. The prospectus proposes a railway from Minehead on the Bristol Channel to the Exe valley and so by Tiverton to Exeter and the English Channel. The writer says:

Since the project of the intended Ship Canal was abandoned, Rail Roads have been tried, and they are found to possess such decided advantages over Canals, that unless in some particular cases, there is little doubt but that they will supersede the latter altogether.

He goes on to use in favour of his railway the same argument about the avoidance of Land's End that had already been so often used by canal promoters, and continues:

It is known that there is to be a Rail Road from London to Bristol: from the latter place to Minehead by sea is about forty-five miles, a distance which a steam vessel, with the tide, will perform in four hours upon the average; so that London, Bristol, Exeter, and all the intermediate places will be brought as much nearer in time, as the difference in time will be which it takes passengers and goods to travel from one place to another on the new line, by steam carriages on the Rail Road and steam packets on the Channel, and that which it now takes to go from the same place to the other same place by common conveyances on ordinary roads.

It is curious that the writer, far-seeing in some respects, should not have seen that the railway from London to Bristol would, when extended to Exeter, provide a passage much more convenient than his own, or that steam packets would take away the main difficulties of the Land's End passage, or again that the opening of the railway from London to Bristol would divert to itself much of the trade between Wales and London which was a large part of the argument for an inter-Channels canal or railway.

Such schemes cropped up again in the railway mania of 1845. There was the West of England Central & Channels Junction Railway, with Isaac Cooke's firm, heavily concerned with speculative transport schemes in Somerset, as secretaries, which was to link with the Bristol & English Channels Connection Railway & Harbour (see p. 57) from Stolford on the Parrett estuary to Bridport, supported by the Bridgwater & Taunton and the Chard Canal companies, whose courses were to be used between Bridgwater, Taunton and Chard, and also the slightly earlier Bristol & English Channels Direct Junction Railway, with Charles Vignoles as consulting engineer, from Watchet via Taunton, Ilminster and Chard to Bridport, apparently promoted by those interested in the North Devon Railway. This scheme used old Ship Canal surveys in spite of the large difference of route, and its supporters claimed that 'The Line has been successively surveyed by Mr. Rennie, Mr. Telford, Mr. Green, and Capt. Nicholls, all Engineers of eminent repute'.[12]

Because this line ran through Taunton, whereas the other did not, the *Taunton Courier* backed it strongly against its canal-sponsored rival, in remarkable prose:

We have nothing to say of other local projects. . . . We would however as a general remark, respectfully suggest to all those to whom special appeals have been made, or are making, in favour of fanciful, or fervidly foolish schemes—that it is incumbent on them, and a most serious duty—involving in its exercise the degree of estimation to which influential individuals may be entitled, not to allow the blandishments of puny interests, or selfish purposes, to outweigh the better considered, perhaps more honestly devised arrangements for the public benefit.[13]

Needless to say, neither of the inter-Channel railways got as far as an Act, but it is interesting to see how in the excitement the old idea of a sea-to-sea connection projected itself almost unchanged into the age of railways and steamships. An editorial of 8 October 1845 in the *Taunton Courier* speaks familiar language:

the frightful loss of lives, and other delays involved in the circumnavigation of the Land's End, are thus avoided, and voyages, hitherto exhausting weeks and months in their performance, and during two-thirds of the year wholly impracticable, are thus superseded by the operation of a few hours transit.

These schemes vanished, yet from time to time optimists still produced plans for a ship canal. There were quite a number of these visionary schemes. I have selected the following because it suggested roughly the same route as the Bridgwater & Taunton and the original Grand Western. It is taken from the *North Devon Herald* of 25 August 1870.

The Proposed Great Western Maritime Ship Canal. Mr. F. A. Owen, the promoter of this enterprise, is a sort of English de Lesseps. He believes his scheme will be ultimately accomplished. The *Mining Journal* says:

'The Great Western Maritime Ship Canal is now attracting considerable attention, and the capital required to construct it (£3,500,000) will no doubt soon be asked for from the public. The success of the Suez Canal has given increased confidence to the promoters in the practicability of their proposition. The construction of the canal would do much to open up a much enlarged market for South Wales coal in all the counties south of the line of the Thames, and would diminish the distance between the South Wales coal-field and the various French ports, avoiding, at the same time, the principal difficulties of the voyage. The canal, as stated in the *Mining Journal* of Nov. 27, 1869, is to be 59 miles long and 21 feet deep; the width being at the bottom 31 feet, and gradually increasing to four times that width at surface. It will extend from Bridgwater Bay in the Bristol Channel to Exmouth on the south coast of Devonshire, and it is estimated that from 4,000,000 to 5,000,000 tons of coal would be taken through annually and a harbour of refuge is to be established in connection with it. The project originated with Telford, in 1825, and though it has not been carried out, owing to want of precedent for so great an enterprise, it is considered to offer very good prospects for commercial success.'

CHAPTER V

The Bridgwater & Taunton Canal

THE ship canal had faded away, but the Bridgwater & Taunton was building: the estimate was £34,135, the engineer James Hollinsworth. It was opened on 3 January 1827,[1] when merchandise arrived by barge at Taunton that had been carried from London to Bridgwater in eight days by the *Hope*, owned by the London & Bridgwater Shipping Co. A crowd, shivering in the cold, watched the craft arriving with flags flying, and listened to the ringing of the church bells.

The Tone Conservators and the canal company now took up hostile attitudes. The former hindered the canal's water supply as far as possible; the latter, as soon as the canal opened, broke down the river bank at Taunton and forcibly connected the canal with it, afterwards sending canal barges on up the river to the wharf at Taunton bridge.

The Conservators now had to compete with an up-to-date adversary, and they were doubtful whether their Acts allowed them to reduce tolls while they had debt unpaid (owed, ironically, to the canal company). Fortified by counsel's opinion they made heavy reductions, from 4s per weigh of 3 tons first to 2s and then to 1s 6d, only to find proceedings taken against them in the Court of King's Bench by the canal company. They won, and the tolls remained down. The canal company then, on 28 August 1827, served notice that they were going to take over the Tone under the powers of the 1811 Act, which were, they conceived, renewed in the 1824 Act. They also re-offered the £10 pa for each parish that had been accepted in 1811. It was now refused, and under the supposed powers of the 1824 Act assessment juries were set up, who assessed the compensation at £300 for each parish. These sums were tendered and again refused, while the Conservators took no notice. In November the canal company forcibly took possession, turned the superintendent, William Goodland, out of his cottage, raised the tolls on the river to 4s again, and on their canal to the Parliamentary maximum, and ceased to maintain the river as a navigation. The Court of King's Bench thereupon ruled that the payments had not been made within the time allowed by the Act, and ordered the canal company to give up the river. They did not do so, and they and the Conservators brought actions against each other in the High Court. Eventually, in February 1830, the Conservators were awarded the river again, with the ruling that the limitation of time in the 1811 Act was compulsory, and that while the canal company could agree with them to give up the river, it could not compel them to do so. They took physical control again in July.

The canal company now began to make soothing noises and money offers, while the Conservators, very cross indeed, cut the tolls of the river that the canal company had raised, kept the canal as short of water as possible, and built a dam to separate it from the river at Firepool Lock below Taunton bridge. When a Chancery Court decision ordered them to remove the dam, they did so, but resolved:

> That Notices be served on all persons, entering with Boats or other Vessels the River Tone from the Bridgwater and Taunton Canal, that actions will be brought against them for so entering, and also that Tolls will be expected of them for using any part of the River for the purposes of navigation.[2]

Another burst of litigation now began, starting with a Bill of Indictment against the Tone's superintendent at Wells Assizes in April 1830, and ending before the Lord Chancellor. There seemed no end. The canal company made a final offer of purchase, which was turned down in November 1831 on the ground 'That the Conservators cannot entertain any proposition which involves in it the abandonment of their duties . . . nor can they consent to a relinquishment of their rights. . . .'[3] The company then proceeded to promote a bill for the compulsory acquisition of the Tone over the heads of the Conservators. These at first became even crosser than before, but eventually they realized that they would do better to negotiate, and agreed terms on the lines of the last offer they had received. The authorizing Act was passed in July 1832. The remaining Tone debt was already held by the canal company, and so was cancelled. The company had to pay £2,000 to the Conservators; settle their small debts amounting to £528 17s 9d and interest, and also their law bills; rebuild two arches of North Town Bridge at Taunton; put the Tone into full navigable order and maintain it so; do certain works on the river; limit their tolls on coal and other goods through Firepool lock and on to the river; build a direct canal between the Tone and the Grand Western Canal; and submit to an annual inspection of the canal by the Conservators to see that it was properly maintained. If it were not, the Conservators had the right to repossess themselves of the river. In return, the Conservators on 13 July 1832: 'resolved that for the considerations aforesaid the Company of Proprietors shall and may forthwith enter into and take possession of the said River Tone and the rights and privileges of the said Conservators.'

The long controversy was over. On 11 September:

The Conservators of the River Tone and a number of other persons connected with the Bridgwater & Taunton Canal yesterday morning embarked in a capacious barge at the junction of the two navigations near this town (Taunton) and proceeded through the line of Canal to the River Parrett near Bridgwater. The attestation of cordiality which now prevails between the parties after the long continued course of litigation which opposing interests have engendered will, we trust and believe, be followed by an early consummation of the full benefits both to this town and elsewhere which have been anticipated from the completion of the Canal.[4]

In 1838 the Conservators applied to the Court of Chancery for an order to dispose of the £2,000 and in 1843 such an order was made to carry out the intentions of the Act of 1699; £1,000 was laid out in building a wing to the Taunton and Somerset Hospital, and £1,000 invested in bonds of the Taunton Market Trust towards its maintenance.

The last years of the Tone Navigation before the canal was opened were prosperous by the standards of the river. In 1823 it carried:

	Tons
Bridgwater to Taunton, coal	28,500
Bridgwater to Taunton, goods	1,879
Bridgwater to Ham Mills, coal	7,385
Bridgwater to Ham Mills, goods	238
Langport* to Ham Mills, goods	712
Taunton to Bridgwater, goods	802
	39,516

The tolls received for that year were £2,194. A Taunton man, writing his reminiscences of this time, said:

* See the account of the Parrett Navigation, p. 83.

CAUTION.

THE Public are cautioned against believing certain statements which are abroad respecting the effects of the CANAL BILL now before Parliament, whereby the Public are told that they will be at the mercy of the Canal Company, to charge them whatever price they please for Coals. THIS IS UTTERLY FALSE, as the *Canal Company's Powers are limited,* and the very utmost Toll they can levy under their Act, is 2 shillings per Ton. Their present Toll is 1 shilling; so that if they were to do the utmost they are empowered to do by Law, it would only be an increase in price of 1 Shilling per Ton, or not quite *three farthings per hundred* above the present price.

TROOD, Printer, Bookbinder, and Auctioneer, TAUNTON.

*Confirms your statement that the Town will lose £5000 [?]
per annum in article of Coal only*

5. & 6. Two broadsheets issued at the height of the dispute between the Tone Conservators and the Bridgwater & Taunton Canal Company

TENDER MERCIES,

INTENDED BY THE

Canal Company.

In answer to a Handbill I have just seen, headed " CAUTION," evidently emanating from the Friends of the Canal, and consequently, the enemies of the People of Taunton, I do assert, that before the Canal Company seized on the Old River, I used to pay *Five Shillings* for every Boat Load of Timber sent from my Wharf at Bathpool; but immediately the River was seized on by the Canal Company, they gradually increased their demands, and ultimately extorted *Ten Shillings* per Boat. God only knows what they would now charge had they not been dispossessed of the River! So much for the Tender Mercies to be expected from the Canal Speculators, who intend to engross to themselves all the Trade of Taunton, and all the benefits to be derived from the River Tone.---Fie on them!!

WM. YATES,

Timber Merchant,

BATHPOOL,

W. TOMS, PRINTER, TAUNTON.

The bridge and its vicinity were fifty years ago the liveliest part of the town, as both coal and goods were brought by water and unloaded there. About one hundred men were employed in this business and were some of the roughest and coarsest lot in the whole town. When the river was low there was a difficulty about getting the barges up to the yards: when that was the case all the boaties (including 'Burpool Tom' and 'The Devil'), united in pulling, cursing, swearing, and thrashing their horses, and would call the boat they were pulling horrid names which would shock the present generation. It was no uncommon occurrence to see them stop in the midst of their pulling and hauling and fight like bulldogs. What with blood and smut they looked like devils incarnate. No one thought it worth their while to interfere or stop them.[5]

In accordance with the Act of 1832, in 1834 a cut was built from the Tone just below French Weir (above North Town bridge at Taunton), with two locks, to the Grand Western Canal at Frieze Hill.*[6] The Tone had to be kept navigable by the canal company. At this time about 13,000 tons of traffic, mainly coal, were worked up to Ham Mills in 12-ton craft for distribution as far south as Ilminster. Almost all of this was transferred to the Chard Canal when the first section from Creech St Michael to Ilminster was opened in 1841. Most of the traffic going higher than Ham, to Taunton or intermediately, transferred to the Bridgwater & Taunton soon after it opened: craft were larger and freight charges much less. What remained came mostly from the Parrett Navigation after the lock at Huntworth was closed in 1841. Craft for Taunton then had to use the Tone, or go down the Parrett and into the canal through the dock.

For the calendar year 1849, Tone tolls were £95 and rents £9, and for 1852, £86 and £8. For the year ending 24 June 1861, tolls were £128 and rents £7, though these were £71 less than the cost of maintaining the river. Tolls for 1865 were down to £68, and for 1876 to £33. In the 1870s there was still some trade to Taunton, but all that on the river averaged 1,841 tons for 1870–76 inclusive. The canal company protected themselves from any compulsion by maintaining the Tone debt in existence, to which they added their own law costs, and then 6 per cent per annum in accordance with the 1699 Act. The debt therefore increased each year, and so they could demonstrate that the revenue was insufficient for maintenance.

* It is uncertain whether the two lines, which met at different levels, were ever physically connected.

The Bridgwater & Taunton had probably cost about £97,000 to build; £57,000 had been raised from shareholders of the company and its predecessor the Bristol & Taunton, and about £40,000 in loans, including £15,000 lent in 1826 by the Exchequer Bill Loan Commissioners. By 1837 this last debt had been reduced to £10,000. It was then paid off by Isaac Cooke and others who were given a first charge on the canal in exchange. Here are the results of its early years:

Y.e. June	Traffic Tons	Gross Revenue £	Net Revenue £
1840	89,235½	7,099	5,126
1841	90,297	7,499	6,011
1842	118,216	11,349	8,239
1843	103,670	10,787	6,691
1844	97,202	9,051	6,303
1845	86,844	7,455	3,363

Its effect upon the port of Bridgwater is shown by the rise in the tonnage handled there from 75,000 tons in 1822 to 113,000 in 1829 and 129,000 in 1834.

The fortunes of all the waterways based on the River Parrett were threatened when in May 1836 an Act was passed for building the Bristol & Exeter Railway with branches to Dunball wharf and Bridgwater. Others to Yeovil and Tiverton were authorized in 1845. The Bridgwater & Taunton reacted vigorously. The prospective opening of the Grand Western, and later the Chard, Canals promised extra traffic, and they had a good base at Bridgwater if it could be made sufficiently secure to withstand railway competition. In 1835 an engineer, H. H. Price, had at the request of Bridgwater corporation reported on the practicability of building a ship canal from Combwich to Bridgwater, a dock there, and extending the Bridgwater & Taunton from its present junction with the Tone at Huntworth to the dock.* The ship canal idea was laid on one side, but in 1837 the canal company obtained an Act[7] to extend their line into the Parrett below Bridgwater, and to build a ship dock at the junction. Thus the port would be improved, because ships had hitherto had to lie at wharves in the river, subject as it was to fast-running tides and the bore as well, while the canal would carry all the traffic that arose from the dock. The Huntworth lock and basin were closed, a mile-long canal

* Such a ship canal had been proposed as far back as 1723, and again by Easton in 1825 and Jessop in 1829.

loop was built by the company's engineer, Thomas Maddicks, round the town of Bridgwater, the dock was excavated, and the whole was opened on 25 March 1841. There was the usual jollification.

The day was ushered in by some merry peals of bells from St Mary's tower, and the firing of cannon near the docks, which aroused the sleepy inhabitants, and caused them to flock near the scene of action at a very early hour, and by high water the number amounted to several thousands, amongst whom were a great number of the fair sex, including the youth and beauty of the town. At a few minutes before eight o'clock a.m., we perceived the *Henry* proceeding at a rapid rate, and at eight o'clock precisely . . . she entered the harbour in a very majestic style, with a band of music on board playing the national anthem, amidst the cheers of the crowd and the firing of cannon, and immediately after, five other vessels graced the harbour. The proprietors gave entertainments at several of the inns. One at the Clarence, . . . included the proprietors, town-council, and a great number of gentry from the town and neighbourhood, specially invited to celebrate the occasion. . . . A dinner was also given to the tradesmen, Captains of vessels, etc. at the George Hotel, and those of the lower class were regaled with good old English fare of roast beef and plum pudding at the smaller inns of the town.[8]

The cost had been about another £100,000, and heavy mortgages were incurred. The first full year's operation of the extended enterprise to 30 June 1842 yielded a revenue of £11,372. Against this had to be set interest charges of £7,376 on the mortgage and loan note debt of £141,300, and costs of management and repairs of £2,300 for the canal and basin and £300 for the Tone. The surplus of £1,396 for the year held out little hope to the shareholders, and was a precarious basis for the canal's side of the competition that immediately began as the railway was opened from Bridgwater to Taunton on 1 July 1842, and tolls were lowered. In August 1844 the Bridgwater & Taunton agreed to pay the Grand Western ½d per ton of coal carried on their canal for every mile up to 14½ that it also moved on the Grand Western; at the same time the two companies agreed toll charges.

In 1844 also the Bristol & Exeter Railway was opened throughout, followed by the Tiverton branch in 1848. That to Bridgwater was not built; instead, Bridgwater corporation built the 'Communication Works', a horse tramway about half a mile long, joining the wharves on the Parrett to the railway; this was opened in

1845. A year earlier the set of wharves at Dunball, farther down the river, had also been joined to the railway by a tramway, privately-built by two directors of the Bristol & Exeter, so that coal could now be sent to the west in competition with the Bridgwater & Taunton and the Grand Western, by the Old Dunball Coal Train. Some shipowners had then diverted their craft from Bridgwater dock to Dunball wharf.

The combination of these circumstances with the railway mania of 1845 made the canal company think that its best course might be to follow the crowd, and abandon the water for the rail. It therefore promoted the Bridgwater & Taunton Canal Railway, a conversion of the canal; with its associate, the Chard Canal Company, the Chard Canal Railway, another conversion; and lastly, the Bristol & English Channels Connection Railway & Harbour, a revival of the old inter-Channel ideas. This was to run from Stolford on the Parrett estuary to Bridport, making use of the two other proposals for railways as part of its line. There were four members of the Chard Canal committee and two of that of the Bridgwater & Taunton on its provisional board; the secretary was John Farquhar, manager of the two canals, and the solicitors, Isaac Cooke & Sons of Bristol, who acted for both, and were also financially interested in them.

Simultaneously with this north–south plan, there was an east–west plan in association with the Kennet & Avon Canal. In 1845 that company were promoting the London, Newbury & Bath Direct Railway. It was now proposed to build a second line of railway from the London, Newbury & Bath at Devizes to Bridgwater, to be called the London, Devizes & Bridgwater Direct, and five members of the Kennet & Avon committee, already directors of the L.N.&B.D., joined the board. From Bridgwater to Taunton the line of the Bridgwater & Taunton Canal Railway would be available as part of an extended scheme called the West of England Central & Channels Junction Railway, which would carry the rails on to near Okehampton, to join the projected Devon & Cornwall Central to Falmouth and Penzance. Here was a somewhat shorter alternative line to the Great Western, Bristol & Exeter, and their extensions, and

> What makes this scheme the favourite in the race is, that meeting at Devizes with the London, Newbury & Bath, and at Bridgwater, with the West of England Central and Channels Junction Railway—both made on Canals the *property* of the respective Companies—the whole distance from London to Oakhampton, will be open to the public at very much lower

Bristol & Taunton
CANAL.

The Inhabitants of the Town of TAUNTON, and of the Parishes of *Taunton Saint Mary Magdalene,* and *Taunton Saint James,* are respectfully informed that a Petition to the

House of Lords, in Favor of the Bill,

Now before Parliament, to Abridge, Vary, Extend, and Improve the BRISTOL and TAUNTON CANAL NAVIGATION, is now Open for their SIGNATURES, at the Offices of Mr. JOHN BUNCOMBE, at Mount, and of the undersigned in Hammet-Street.

The Bill, after an Investigation of Seven Days, and a patient hearing of Witnesses on both sides, has passed through a Committee of the House of Commons, by a

Majority of Twenty-Six to Six,

of Members present on that Committee.

HENRY JAMES LEIGH,
Agent for the Canal Company at Taunton.

TAUNTON, 25th May, 1824.

fares, and, as the three companies will be making their lines at the same time, several years sooner than the same accommodation can be supplied by any other party.[9]

The London–Devizes–Bridgwater–Okehampton–Penzance plan fell through with the collapse of the Kennet & Avon's projected railway, and was briefly succeeded by the Bath, Wells & Exeter Railway, which proposed to build a line from near Limpley Stoke via Bath, Wells, Glastonbury, Langport, Ilminster and Chard to join the Exeter & Bridport projected line, with a branch to Bridgwater. This company proposed to arrange with the Somersetshire Coal Canal and Parrett Navigation proprietors for the conversion of part of their lines.

The plan to convert the Bridgwater & Taunton Canal into a railway was opposed by the Tone Conservators, who threatened to exercise their power to reopen the river were the canal to be closed. It then became a proposal to carry out a small part of the inter-Channel scheme, the building of a line from the Bristol & Exeter Railway at Bridgwater to Stolford, and a harbour there. In this form, and including power to stop up the canal, it was authorized in 1846, the company's name being changed to the Bridgwater & Taunton Canal & Stolford Railway & Harbour Company. No action to build the line followed.

John Farquhar, the canal's manager, had from 1843 also acted as a carrier. With George Cooke of the company's firm of Bristol solicitors, he had formed the Bridgwater & Chard Coal Company to trade both in coal and general merchandise, and hired vessels for the Bristol Channel trade, being guaranteed against loss by the company. In 1845, however, Thomas Reynolds, the canal's chairman, took over on behalf of the second mortgagees, and Farquhar was also appointed receiver. The agreement formerly made with the Grand Western was now ignored, with the latter company claiming £403 for its first year's working. Thus arose a quarrel between the companies; partly as a result, the Grand Western and the railway reached agreement in 1848, by which the Bridgwater & Taunton lost considerable trade.

At the end of 1851 the company listed its debts, nearly all mortgages, as £118,130. Its results are shown on the following page.

In 1848 the Bristol & Exeter Railway had made a carrying agreement with the Grand Western Company that virtually confined the Tiverton trade to those companies, to the exclusion of the Bridgwater & Taunton (see p. 112). About the end of 1849, however, an official of the railway, Mr Harriott, seems to have thought a better policy would be to end competition by both canal com-

panies, and in June 1850 he approached the chairman of the Bridgwater & Taunton, Thomas Reynolds, with a suggestion that the
railway might be interested in a proposal from his directors. Reynolds, also a principal mortgagee, then wrote to the railway board,
only to be told that

> this Board has not authorized any proposal to be made to the
> Bridgwater & Taunton Canal Company, but that the Board is
> ready to receive any proposals that the Committee of the Canal
> Company may desire to make to this Company.[10]

Y.e. 30 June	Traffic Tons	Gross Revenue £	Net Revenue £
1846	73,440½	5,637	3,226
1847	79,613	6,844	3,091
1848	73,306	6,454	3,181
1849	58,407	5,326	2,503
1850	58,947	5,336	2,428
1851	54,681	4,800	2,774
1852	59,806	4,605	2,715

Calendar Years

1853		4,932	
1854		5,723	
1855		5,008	

By the early 1850s, the traffic pattern had changed. North-
country coal had been reduced in price, and was competing successfully with canal-borne coal at south coast ports; limestone and
culm tonnages were falling because guano and artificial fertilizers
were replacing them, and towns like Honiton drew their supplies
from railway stations and not from the canal head at Chard.

John Farquhar, trying for an offer, now came to a railway
board meeting to be told that a written proposal from his directors
would receive their immediate attention. He lobbied some of the
railway board, but nothing happened. Then, early in 1851, the
railway insisted on a new and much less favourable rates agreement with the Grand Western. This led to a severe drop in that
canal's receipts, and consequently a decision to join the Bridgwater
& Taunton in fighting the railway. Serious rate cutting began, but
not for long. As earlier the Grand Western had gone behind the
backs of the Bridgwater & Taunton, so now, in August 1853, the
latter company proposed an alliance with the railway against the
Grand Western.

Its genesis may be guessed from the fact that early in 1853,

Daniel Fripp (brother of William Fripp, a director of the Bristol & Exeter), a mortgagee of the Bridgwater & Taunton for £10,000, and also of the Chard Canal, applied for the replacement of John Farquhar as receiver of the Bridgwater & Taunton, on the grounds that he carried on a coal-carrying and trading business; also for a receiver to be appointed to take over from him as manager of the Chard on the same grounds. At this time the Bridgwater & Taunton was paying full interest on its first mortgages, but only about half that on its other loans.

His move followed the rejection by Farquhar and George Cooke, his partner in the coal business, of the Bristol firm of solicitors who had a good deal of the railway's overtures, and their determination to keep canal competition going unless the railway offered a purchase price of £8,000 per annum. Fripp wanted, he said, friendly relations with the railway, 'whereby the violent competition at present existing . . . may be terminated', and by nominating J. C. Wall, the chief traffic superintendent of the Bristol & Exeter Railway, who had once worked for a carrying firm on the Staffordshire & Worcestershire Canal, as receiver of both canals, hoped 'by vesting the management of the said Canal and Railway in one person the carrying out of such arrangements as may be deemed advisable respecting the Traffic will be greatly facilitated'. The railway's solicitors added that 'the Canals have been a source of trouble and annoyance to them since the line opened'.

But Fripp had showed too much personal enmity to Farquhar and Reynolds, for his views to be taken seriously. He had written letters addressed to 'The Swindling Bridgwater & Taunton Canal Company', and saying that 'The offence is, that not having the fear of God but instigated by the Devil the Bridgwater & Taunton Canal Company obtained from me under false pretences Ten thousand pounds'. The judge considered that Farquhar appeared to have been an active and useful servant of the canal companies, and if something like the establishment of the coal company had not been done, there was great risk of loss to the canal proprietors. Mr. Farquhar was, however, as respected his duties and his interest in the coal company, placed in an unfortunate position. . . . He thought Mr. Fripp was entitled to have an independent person appointed, and that the Court ought not to refuse this relief because it appeared, or was likely, that Mr. Fripp was connected with the Bristol & Exeter Railway Company.[11]

Wall was not accepted as a suitable receiver, and C. W. Loveridge of Chard, a partner in Stuckey's Bank, was appointed. The Bridgwater & Taunton (and also the Chard, for which Mr Fripp had Loveridge appointed receiver the same day) now clearly passed under railway influence. It was soon shown.

The proposal made in August by the Bridgwater & Taunton is recorded as follows by the railway company:

The Deputy Chairman reported that the Committee on Canals . . . had on the 22nd. inst. an interview with a deputation from the Bridgwater & Taunton Canal Company (consisting of Mr. Reynolds, Mr. Cave and Mr. Paul) and received from them a written memorandum, an arrangement which the deputation intended to propose to this Company after giving notice to the Grand Western Company of their intention of so-doing.

Read the Memorandum alluded to, of which the following is a copy:

'The Tolls and Charges to be levied and received from the Bristol and Exeter Railway Company on all articles hereafter to be imported into or exported from the Port of Bridgwater to be such as shall practically ensure the Conveyance of all those articles on the Bridgwater and Taunton Canal at the full Parliamentary Toll to and from all places within seven miles of the Taunton Station of the Bristol and Exeter Railway, (except Wellington) and also the local traffic between the towns of Bridgwater and Taunton, and with regard to Coal and Coke, whether imported into Bridgwater or brought from any other place.

'The Siding from the Taunton Station to the Grand Western Canal to be removed and not reinstated, and no other Station or Siding to be made between Taunton and Wellington, or between Taunton and the branch railway to Yeovil.

'The Bridgwater and Taunton Canal Company to charge and receive their full Parliamentary Toll on all articles conveyed upon their Canal imported into or exported from the Port of Bridgwater and which shall also be conveyed upon the Grand Western Canal.

'The arrangement to be made for seven years.'[12]

The Grand Western saw that the end had come, and the same railway board meeting in October 1853 that approved the arrangement with the Bridgwater & Taunton saw the Grand Western's superintendent arrive with the company's proposal for a railway lease, which was agreed.

So matters rested for some years. Then in late 1863, seemingly to forestall possible LSWR encroachment, the railway company moved to buy the Grand Western, Bridgwater & Taunton and Chard Canals. The first was purchased in 1864, the portion of its line between Taunton and Lowdwells being closed in 1867, so breaking the through route to Tiverton; the others in 1866 and 1867. The Bridgwater & Taunton was bought for £64,000, the receiver being discharged early in April 1867 and the railway obtaining possession on the 8th. At this time there was a first mortgage of £10,000 and second mortgages of £89,992 on the canal, all bearing interest at 5 per cent. The tolls, averaged over the five years ending 30 June 1865, had only paid interest on the first mortgage and about 2¼ per cent on the second mortgage debt. Out of the purchase price the first mortgagee got his £10,000, the second mortgagees £50,625, and the shareholders what little remained after debts had been paid. One of the petitioners against the bill saw clearly that there had for some time been an agreement between waterway and rail. Clause 14 of the petition of the Feoffees of Meredith and Ackland's Charity at Taunton stated:

That the course of the Bristol and Exeter Railway between the said Towns of Bridgwater and Taunton runs nearly parallel to the said Canal, and that the charges and tolls now taken by the said Railway Company for conveyance of heavy goods along the said Railway are carefully adjusted, so that the tonnage and truck hire on the Railway are just equivalent to the tonnage and expenses of shipment on the said Canal.

When the Act was passed the Conservators of the River Tone were apparently pleased, though maybe desirous also of conveying a warning:

A print of the Bridgwater and Taunton Canal Act, 1866, which received the Royal Assent on the 28th. ulto. having been laid before the meeting by the Treasurer, the Conservators desire to record their satisfaction at finding that the Canal has been thereby transferred into the hands of a Company who have the means as well as the will, as expressed in the 23rd. sect. of the Act, of maintaining and keeping the navigation in thorough and complete repair & efficiency in all respects, so that it may always afford a good and sufficient water communication between the Towns of Bridgwater & Taunton; to secure which object has ever been the earnest and zealous endeavour of the Conservators.[13]

The purchase Act was stated not to be an abandonment of the Tone within the meaning of the 1832 Act, which set out the conditions upon which the Conservators could resume the river. But the railway company undertook only to maintain the canal, together with the towing path beside the river from North Town bridge, Taunton, to the gas company's wharf.

Meanwhile the railway company had in 1865 built a new timber landing stage fitted with a steam crane on the Parrett at Bridgwater to transfer coal from vessels in the river to the 'Communication Works', the old horse tramway from their line to the older Bridgwater wharves which they had leased from 25 December 1859, and which they now decided to convert to a locomotive branch and extend across the river to the canal dock. In November 1867 this was opened to the wharf, itself soon afterwards extended and given another steam crane, but the bridge to the dock with its telescopic span was not finished until March 1871. Thereafter the line was worked partly by locomotive and partly by horsepower. Having bought the canal, the company closed the dock for alterations, repairs and new gates. It was reopened on 1 January 1868. In 1867, however, they had been authorized to buy and extend Dunball wharf, lower down the river where it was deeper, and make a locomotive branch to it to replace the tramway. This was completed in November 1869, with a wharf extension finished in 1874. Dunball's competition must have affected the trade of the dock as well as of the steam crane wharf.

From 1873 onwards the dock took about £1,000 pa in dues, and handled about 130,000 tons of traffic annually, mostly imports of coal and timber. Dunball wharves dealt with about another 100,000 tons a year between 1876 and 1881, after which quantities fell somewhat. The opening of the Severn tunnel late in 1886 greatly affected the trade of the dock and wharves, for much of the south Wales trade that had formerly been by sea was now transferred to rail.

In 1869 the Bristol & Exeter Company renewed a number of lock gates and bridges on the canal, and in 1871 added to the dock warehouses.

They worked both dock and canal hard, and maintained its receipts. The figures appear at the top of next page.

In 1876, however, the Bristol & Exeter amalgamated with the Great Western, who had very different ideas about canals. By 1890 the canal tonnage was 13,809, all of it coal and timber. From 1870 onwards the Conservators from time to time complained of the canal's condition. Between 1896 and 1901 there were claims against the railway company on the grounds of short water, and the company therefore offered to carry temporarily by rail at canal rates. Pumping power at Creech was then increased, but the traders who had transferred then stayed on rail. By 1905 the canal

V. Chard Canal: (*above*) Incline-keeper's house at Crimson Hill; (*below*) Ilminster inclined plane descends the hill to the canal on the left

VI. Glastonbury Canal: (*above*) Remains of Shapwick lock; (*below*) the abutments of the aqueduct at Ashcott, with the Somerset & Dorset line beyond, in 1966

Year	Total Receipts inc. Dock Dues £	Canal Tolls Only £
1868	3,298	
1869	3,206	
1870	3,310	2,563
1871	2,284	2,213
1872	3,135	1,901
1873	3,612*	1,831
1874	3,364*	1,709
1875	3,338*	1,616
1876		1,629

In 1868, 30,018 tons were carried, and an average of 23,336 tons for the years 1870–76 inclusive.

tonnage was 6,420; there was none in 1907. The figures of receipts (excluding dock dues) for the years 1878 to 1902, averaged over three-year periods, are as follows:

Years	Receipts (exc. dock dues) £
1878–80	1,664
1881–3	1,607
1884–6	1,535
1887–9	1,459
1890–2	1,275

Thereafter the canal remained virtually unused. During the Second World War the military removed twelve swing bridges and replaced them with fixed bridges, some of which have remained.

With the closing of the Parrett Navigation about 1878, little use remained for the Tone, except for the short section above Firepool, and probably traffic above Ham then ended. The last barge used the Burrow bridge—Ham section about 1929. The river remained in theory navigable until 1967, when, with the consent of the British Waterways Board, the Minister of Agriculture agreed to the application of the Somerset River Authority to extinguish the navigation rights below Firepool, and repeal the relevant Acts under s. 41 of the Land Drainage Act, 1930, so that a weir could be built there as part of a flood relief scheme for the Tone, which involved straightening the whole river from French Weir in Taunton to Burrow Firepool bridge. Above the Weir the river still remains a navigation for a short distance.

* These figures include about £360 of Grand Western receipts also.

E

CHAPTER VI

The Chard Canal

CHARD might have been served by an inter-Channel waterway; this possibility had been there from 1769, when Whitworth had made his survey, to the Ship Canal Act of 1825. These failed, but the opening of the Bridgwater & Taunton Canal in 1827 led Chard people to wonder whether they could not be linked with it. A Mr W. Hanning, who had interests in the Westmoor area about 4 miles from Ilminster towards Langport, had in 1831 given notice of his intended canalization of the Westmoor main drain to the Parrett as part of a local enclosure scheme. He then suggested that the Bridgwater & Taunton Company should make it, and so improve communications by water towards Ilminster and Chard. James Green, who was engineer to the enclosure, examined the Westmoor scheme, and suggested that it would be more sensible for the canal company to make a branch direct from their canal to Chard. The Bristol backers of the Bridgwater & Taunton considered that, should the Westmoor scheme be carried out by someone else, they would lose £1,200 pa in tolls by diversion of traffic at Bridgwater. Understanding Hanning to be willing to withdraw his scheme (he died in 1834), the Bristol backers expressed interest if Chard people would subscribe £5,000, and Green was employed to do a detailed survey. In carrying it out he was clearly influenced by his own inter-Channel project of 1822, and by the work he was then doing on the Grand Western. He proposed a canal of five pounds, to leave the Bridgwater & Taunton at Creech St Michael, pass the three intervening ranges of hills, each higher than the last, by lifts or inclined planes, and tunnels where necessary, to end at Chard, 231 ft higher and 13½ miles from the junction.

He proposed two lifts at Thornfalcon (28 ft) and Wrantage (31 ft), and two inclined planes at Ilminster (94 ft at 1 in 6) and Chard Common (78 ft at 1 in 12), with tunnels at Lillesdon and Crimson Hill. His estimate of money was £57,000; of time, five years.

The Bridgwater & Taunton Company accepted the survey, and supported the formation of a Chard Canal Company and the promotion of a bill. By reckoning that the whole existing needs of the inhabitants along the line for coal and merchandise would be met by the canal, as well as upon a new trade in stone, they thought a net revenue of £6,000 pa could be got on a tonnage of 78,940, to yield 10½ per cent to the shareholders and £2,500 pa extra to the tolls of the Bridgwater & Taunton.

The scheme was very largely the creation of the five principal men, all from Bristol and its neighbourhood, who were behind the Bridgwater & Taunton Company: Isaac Cooke whose firm of solicitors had helped to promote the old Bristol & Taunton, John and William Cave, Joseph Reynolds, and Joseph Cookson, chairman of the Severn & Wye tramroad company and presumably interested in selling Forest coal to Somerset residents. Out of the £46,850 that had been subscribed when the bill was sought, they had put up £40,000 in equal shares. Of the fifty-six other names on the subscription deed,[1] one (also a Bristolian, James Fripp), put his name down for forty shares of £50, one for twenty, one for ten, two for five, and six for two; the other forty-five, mainly Chard small tradesmen, for one each. One of those with single shares, described as a 'gentleman', could not write, and made his mark. The borough and parish of Chard, with its population of some 5,500, provided only £3,250 of the subscription. Clearly local enthusiasm was minimal, and doubt considerable, for the Act[2] of June 1834 provided that until the full authorized capital had been subscribed, land could not be bought compulsorily. This was £57,000, with £20,000 more if necessary, seven years being allowed for construction.

By this time Green was no longer engineer, due presumably to his failure with the Grand Western. One result was to eliminate the lifts at Thornfalcon and Wrantage, presumably because those being built on the Grand Western having teething troubles, and to substitute inclined planes. William Cubitt was consulted, and recommended a man of only twenty-two, Sydney Hall. He had been a pupil first of George Edwards of Lowestoft and then of Cubitt, before working for two years at the Horseley Bridge ironworks. The Chard Canal was his first job. The assistant engineer was Isaac Whitewood.

The Act, in the interests of the landowners, had laid it down that until half of the tunnels at Lillesdon and Crimson Hill had been completed, no part of the line more than 2 miles from the far side of Crimson Hill might be begun without the special consent of certain landowners, a stipulation that delayed the work. On Midsummer Day, 1835, cutting began at Wrantage, but it was not until the autumn of 1837 that work was so far forward on the tunnels that construction could begin on the Chard side of Ashill, 2 miles beyond Crimson Hill. Tunnelling was being carried on day and night at Crimson Hill, for on 17 October 1837, a labourer was killed at midnight by a fall of earth. The Bristol & Exeter Railway was authorized in 1836: its line passed under the Chard Canal at Creech in an invert. About this time pessimism set in about the prospect of the canal ever being finished, and about its future, which never afterwards entirely died away, and in the following year a visit of shareholders to the workings was arranged. The Chard reservoir was now begun, and in 1839 the Lillesdon and Crimson Hill tunnels were finished.

At about this time, after the ground had been cut to a depth of about 10 ft above the line of the later tunnel at Ilminster, the engineer seems to have decided that Green's original line involved an unnecessarily large embankment at Dowlish Ford and excessive work all the way to Chard Common incline. He therefore lowered the line considerably by inserting Ilminster tunnel at the head of the inclined plane, then lifting it about 7 ft by a lock 1 mile southwest of Dowlish Ford, and also increasing the height of Chard Common incline.

In March 1840 an Act gave power to raise another £80,000 by preference shares, and £26,000 by mortgage. By August 992 of the 1,140 new shares created had been taken up and calls allowed work to continue. In 1841 another Act extended the seven-year limit.

In August 1840 shareholders were told that the exceptional winter rains had delayed work, which since then had been going well. In September a number of hopeful Chard inhabitants 'applied for spirit licences for houses situated near the termination of the Canal, in anticipation no doubt of a thriving trade when the works are completed'.[3] They were all refused by unsympathetic Justices, until in September 1841 one licence was granted for the Furnham Hotel near the basin and the wharves. At Chard in early 1841,

> the entrance to the town from the Ilminster road now presents a very busy and altered scene from the active operations employed in the erection of warehouses and wharfs for the canal company at the basin; these are being constructed in masonry of a solid and substantial nature, by Mr. Henry England, of Chard.[4]

On 15 May 1841 the canal was formally opened as far as the coal, culm and merchandise wharf Summers & Slater had established at Ilminster:

the directors, accompanied by their friends, entered the Canal at Creech St. Michael from the Bridgwater & Taunton Canal in a boat appropriately fitted up for the occasion, and proceeded along the line towards Ilminster amidst the cheers of a large concourse of spectators. Several boats laden with coal and culm followed, belonging to Messrs. Summers & Slater, and other merchants who have taken wharfs adjoining the Canal.[5]

The *Taunton Courier*, after thus describing the event, says that the two intervening planes failed to work, and the procession never got to Ilminster. But the *Sherborne Journal*, in an issue published on opening day, says: 'towns . . . may obtain their coal etc. at Ilminster . . . even in the present unfinished state of the canal works. There yet remain to be brought into operation the several inclined planes between Chard and Creech'. It seems more likely, therefore, that the canal company was for the time being providing land transport round Thornfalcon and Wrantage planes.

Thornfalcon was reported at work on 27 July: 'nothing could exceed the facility with which the boats loaded with coal were drawn up on Tuesday. The moving power on the plane consists in cassoons, the empty boat in the cassoon full of water pulling up the loaded boat in the empty cassoon.'[6] Wrantage plane must have been ready at the same time, and regular trade to Ilminster began.

Meanwhile the great reservoir at Chard had been built, where in the original Act Lord Poulett, as lord of the manor, had taken power to fish and to have boats. The local newspaper let itself go on the subject:

The reservoir has risen considerably during the past week and continues to present new features of interest and attraction as it is viewed from various points in the neighbourhood—its broad expanse of surface of the deepest blue, resting sometimes in peaceful repose, and at others ruffled . . . and tossed to and fro, covered with crested waves of no mean size, and its shores washed by breakers, and besprinkled with angry foam.

The paper then went on to ask for a suggestion for a

name for this fine sheet of water, one which shall be in classically correct taste, consistent with the nature of things (as the schoolmen say) and yet perfectly intelligible to, and adapted to the articulating organs of, the *profanum vulgus*.[7]

One is not surprised that the paper soon afterwards ceased publication.

By October the wharfs and warehouses at Chard were finished. As 1842 began, the company were busy getting the next section ready to Dowlish Ford. On the Ilminster inclined plane 'upwards of two hundred hands have been employed day and night, presenting a novel and singular appearance, the whole length of the incline being brilliantly illuminated with fires and lights, to enable the workmen to proceed'. Then, on Thursday, 3 February,

> some boats which had been laden with coals, in waiting, were attached to the machinery, which being set in motion, ascended majestically up the incline. . . . Crowds of people were assembled to witness the sight; the moment for starting was the signal for a general cheer. The Ilminster band . . . struck up one of their best tunes . . . on arriving at the top of the incline . . . the band then took their station in a boat . . . and proceeded through the tunnel, playing the national air and rule Britannia, assisted by some first-rate vocal performers, until they arrived at Mr. G. Lang's wharf at Dowlishford.

Mr Lang having provided a hogshead of cider, the crowd happily drank it, burst into country dancing, cheered the Queen, and eventually returned to Ilminster, headed by the band, to be greeted by the church bells. It must have been a most satisfactory day, given that the season was early February.[8] The line was further opened to Cricket Bridge in April.

Meanwhile Sydney Hall was working on Chard Common incline. Toms & Sons had taken a wharf at Chard, where they proposed to stock coal, culm, building materials, salt, soda and other goods, and to distribute them to Crewkerne, Axminster, Colyton, Honiton, Beaminster, Lyme Regis and elsewhere. There was some delay when a rope broke at Wrantage plane, damaging both caissons; there was also some consternation, for the day before a party of ladies had been given a ride on it. Delay also from the construction of the Bristol & Exeter Railway's invert under the canal at Creech. But at last, on Tuesday, 24 May, the canal was open, except for the authorized branch from Chard to the Chaffcombe road, which was not built. The tendentious *Woolmer's Exeter & Plymouth Gazette* reported it thus:

> A few lines will suffice to announce the opening of this canal. This anxiously looked for event took place, quite accidentally, on the Queen's birthday. We say accidentally because, with the uncertain machinery at the inclined plane, and the celebrated frail rope, it was deemed a moral impossibility to say on what precise day any cargo could be brought to the wharfs. Happily, however, with only one more snapping of the rope, a barge

laden with coals was got over the plane on Tuesday evening, a circumstance, which, although long ago promised, and expected, created no little surprise to the inhabitants, who regarded it in the light of a phenomenon, the 'opening' having been almost entirely given up. We have not the task of recording a public dinner on this momentous occasion, nor is it probable we shall, for the matter will hardly bear the ordeal of a public discussion. The actual opening and celebration lasted but a few short hours. The ringers (who were about to let down the bells after ringing for Her Majesty) gave an additional peal or two by desire, and the proprietors of the two wharfs already taken distributed a hogshead of cider to such as pleased to drink it, the finishing of which concluded the evening's entertainment.[9]

But the thing was finished, and on Friday 'the first return train of boats from Messrs Toms & Sons' wharf was freighted with wool for the North of England'.[10] Coal and culm prices fell. Locals were clear that the canal would benefit them: as for the shareholders, 'the income to be derived by the shareholders is not a question for the public generally to consider'.[11] By July 'a new wire-rope, of substantial dimensions, has been substituted for that which has frequently broken at the Chard plane, and Mr. Hall, the engineer, has left. The lime-kilns and coal-yards throughout the line are in active trade.'[12]

The Chard Canal was built for small tub-boats 26 ft by $6\frac{1}{2}$ ft, which alone could use the planes. The three inclines at Thornfalcon, Wrantage and Ilminster were double; that is, they had two lines of railway, on each of which ran a six-wheeled caisson $28\frac{1}{2}$ ft by $6\frac{3}{4}$ ft, filled with water, the caissons being joined together by a chain passing round a horizontal drum at the top. The boats floated in the caissons, motion being obtained by adding water to the upper caisson till it overbalanced that at the bottom of the incline. The plane at Chard Common, however, had a single line of rails (the only such canal plane in Britain) carrying a four-wheeled cradle, on which the boat was carried dry. The motive power was described as 'Whitelaw's patent water-mill'. This was one of Whitelaw & Stirrat's water turbines, which worked on a head of 25 ft and a flow of 725 cu ft of water a minute.[13] Apart from the lock at Dowlish Ford, there was a stop-lock where the canal joined the Bridgwater & Taunton at Creech St Michael. The tunnels at Lillesdon and Crimson Hill were narrow, but that at Ilminster 14 ft wide, so that boats could pass in it.

The amount raised from shareholders was £96,175, and in mortgages £35,000. In 1853 the company also owed £13,623, a few thousands of which probably date from their opening. The likely cost was therefore about £140,000, against the original estimate of £57,000, and from the start the company was unable to pay the full interest on the mortgage debt.

The first year seems to have gone quite well. A new and good business grew up at Toms' wharf in guano, and in August 1843 the company's annual report 'presented a satisfactory statement of the amount of tonnage on the canal, during this its first year; there is also every probability of increased traffic. The committee came up the line in their iron boat to Chard wharf, examining the works throughout'.[14]

But in July 1842 the Bristol & Exeter Railway had been finished to Taunton, whence land carriage to Ilminster and Chard was easy enough. The Westport Canal to the Parrett was by now also competing seriously for Chard and Ilminster traffic, higher land carriage costs being balanced by lower tolls. By 1844 six of the twelve original traders had left the canal for other means of transport. To counter railway competition and maintain a trade on the Chard and the Bridgwater & Taunton Canals, John Farquhar, a Bridgwater merchant who was manager of both, and George Cooke of the Bristol firm of solicitors who handled the company's affairs, formed the Bridgwater & Chard Coal Company, to act as carriers and to trade on the canal. They took over the wharfs of those who did not wish to continue, such as Whiteman & Nobbs and Toms & Sons of Chard, George Toms of the latter becoming the coal company's agent. Absence of competition, however, in 1845, led one firm, Riste & Hill, later Wheatley & Riste, lace manufacturers of Chard, to put eight barges on the canal to supply their own factory with coal and as general merchants. Farquhar seems to have got on their wrong side, and made them think that the canal management, who were also the coal company, were discriminating against them. Stories were then put about that Riste & Hill were being hindered from trading.

For the canal's first three full years, the traffic was as follows:

	Coal Tons	Culm Tons	Other Tons	Total Tons
1843	12,074½	5,237¼	8,523	25,834¾
1844	13,710¼	6,929¾	11,429¾	32,069¾
1845	15,742	7,619½	9,922¾	33,284¼

Railways increasingly threatened. Only four years after the canal had opened, the company sought and obtained an Act[15] to convert part of the line from the Bristol & Exeter at Creech to the Taunton–Ilminster road at Ilminster into a railway, by using the company's unissued capital. They were given five years, but might not close the canal until they had paid off their £35,000 mortgage debt. The railway's three promoters were John Farquhar, Thomas Reynolds of Westbury-on-Trym, chairman of the Bridgwater & Taunton Canal and son of Joseph Reynolds, one of the original promoters, and Isaac Cooke the Bristol solicitor.

In the following year of 1847 another Act[16] extended these plans by authorizing the company, whose name was now changed to the Chard Railway Company, to extend the railway from Ilminster to Chard. By now they had £46,000 of mortgages. These had to be paid off before they could start, and a good deal of ordinary capital had then to be raised before they could re-mortgage themselves. The Bridgwater & Taunton Company were given powers to subscribe, or to buy or lease, or the two companies could amalgamate by treating their shares as of equal value. In 1849 the company had a staff of twelve, with a wages bill of £628 pa. John Farquhar, the manager, was paid £200 pa of this. About this time the second mortgagees took over control of the canal from the shareholders, though they did not put in a receiver.

Nothing happened until 1853. Now a third Act was passed[17] which again authorized the lines of 1846 and 1847, but extended them from Ruishton near Creech into Taunton itself. This Act was promoted by the mortgagees, of whom Daniel Fripp, Joseph Reynolds and Cooke's the solicitors were three of the largest, and was an effort to make their investments worth something by converting what they owned to a possibly productive railway. Peter Bruff was engaged as engineer.

It was a tricky moment, for Daniel Fripp, whose brother William, a director of the Bristol & Exeter Railway, was a first mortgagee for £20,000 with interest some £5,000 or £6,000 in arrear (oddly, William was not critical of the canal management) had sought the appointment of a receiver in the course of a High Court action he had brought against George Cooke and John Farquhar for mismanaging the company to exclude independent traders, so diminishing the canal's trade. This was a continuation of the trouble Riste & Hill had earlier had with the management and with the Bridgwater & Chard Coal Company. Fripp alleged that the latter were virtually monopoly carriers, for coal from south Wales or the Forest of Dean via Bridgwater was the principal traffic, and since only Riste & Hill had not been bought out or given up their wharfs, the canal was run in the interests of the coal company. Low tolls were charged to benefit the latter, though the annual accounts showed losses and only small sums were paid to the mortagees.

The Bristol & Exeter Railway had approached those connected with the Bridgwater & Taunton and Chard Canals to see whether competition could not be ended, and had offered purchase terms which would have given the mortgagees two-thirds of their money back on the former and half on the latter, in exchange for the canals, and Fripp's move—through his brother connected with the railway company—was because these approaches had been ignored by Cooke and Farquhar.

Fripp attained his object, though the court in appointing a receiver upheld what Farquhar, Cooke and the coal company had done as the only possible course in the circumstances. They refused to appoint Fripp's nominee, J. C. Wall, chief traffic superintendent of the Bristol & Exeter Railway, but on 29 June chose instead C. W. Loveridge, a partner in Stuckey's Bank. In August the Bridgwater & Taunton Company made proposals for an alliance with the railway (see p. 62), which were accepted. They ended price cutting, and must have helped the Chard company also by allowing tolls to rise. Receipts from tolls were £1,915 for 1855 and £2,072 for 1856, which seems to confirm this view.

A Chard wharf-book for June–July 1858[18] shows how widely traffic brought by the canal was distributed: customers were served from Axmouth, Axminster, Lyme Regis, Burstock, Sidmouth, Colyton, Broadwindsor, Bridport and Whitchurch among other places. Welsh coal was described as Merthyr, Cardiff or Newport; Forest coal came from Lydney and Bullo Pill; culm for lime-burning from Neath. Other goods handled were fertilizers, salt, soda, bricks, slates, timber, stone and pipes.

For 1855 the total tonnage carried was 23,653¾ tons, all of which moved towards Chard except for 2,179¼ tons of back carriage. In the following year the figures were 25,168¾ tons and 2,602¼ tons. The principal cargoes for 1856 were:

	Tons
Coal, coke and culm	16,823¾
Stone	3,768¼
Grain	1,324
Wool	355¾
Manure	326
Bricks	307¼
Salt	296½
Tiles	273¼
Miscellaneous	1,694

In August 1856 a meeting was held in Chard to advocate railway communications with Taunton, and proposed to ask both the Bristol & Exeter and the London & South Western Railway companies whether they were willing to make the line. There was now some discussion inside the Bristol & Exeter's office upon the possibility that the South Western might build a branch to Chard, buy the Chard Canal and the Bridgwater & Taunton, interfere with the Grand Western lease, and so gain access to Bridgwater, Taunton, Tiverton, and other Bristol & Exeter territory. No move was made until 1859, when at another meeting in Chard a resolution was carried:

> that in the opinion of this meeting a Railway from the London and South Western Railway to Chard with a tramroad to the Canal Basin is of very great importance to the prosperity of the Town, and this Meeting pledges itself to afford the project every support in its power.[19]

By May 1860 this new Chard Railway Company had got their Act. Early in 1861 another bill was presented by an independent company for a railway from Taunton to Chard. It was at first opposed by the Bristol & Exeter, presumably because South Western influence was feared, but after the Act was passed the opposition changed to a successful effort to control the company, which was vested in them from 8 June 1863. The B&ER completed the line as part of their system and opened it on 11 September 1866. Meanwhile the Chard Railway had been bought by the South Western on its opening in May 1863, and to prevent any possibility of encroachment by that company, the Bristol & Exeter board resolved in August to negotiate for the Chard and the Bridgwater & Taunton Canals, and soon afterwards to end the lease of the Grand Western and buy outright.

The Grand Western was bought in 1864, the Bridgwater & Taunton in 1866. A Bristol & Exeter Act of 1867 provided for the closing of the Chard Canal, which had been bought for £5,945. The receiver was not discharged until February 1868, and it is probable that the canal closed then. The reservoir was sold to Lord Poulett, in 1869 the incline machinery disposed of for £745, and in the same year the aqueduct over the main line at Creech was removed, though the invert was not taken away and the rails raised to surface level till 1895.

Today, the remains of the canal still provide much of interest for the enthusiast and the industrial archaeologist. In spite of demolition and tipping, all four inclines can still be visited, all three tunnels are visible, and the Tone aqueduct still stands.

CHAPTER VII

Other Somerset Waterways

❋❋❋❋❋❋❋❋❋❋❋❋❋❋❋❋❋❋❋❋❋❋❋❋❋❋❋❋❋❋❋❋❋❋◆❋❋❋❋❋❋❋❋❋❋❋❋❋❋❋❋❋❋❋❋❋❋❋❋❋❋❋❋❋❋❋❋❋❋

The Rivers Axe and Brue, and the Pillrow Cut

THE River Axe* rises near Wells, and runs past Axbridge to the sea west of Weston-super-Mare. The Brue flows past Glastonbury to the sea at Highbridge in the estuary of the River Parrett.

Richard I gave permission for a port and borough to be made at Radcliffe, now Rackley, to which point it was tidal, by the Bishop of Wells, partly to ship lead from the Mendip mines. Small craft could work higher up to Panborough and Bleadney. The abbots of Glastonbury had their own port lower down, at Rooksbridge near East Brent, on a tidal pill of the old river, whence the Pillrow Cut ran for some 6 miles south across the moor to Mark, and then south-eastwards to join the Brue opposite Burtle. Thence goods were taken up the Brue, through Meare Pool, and so to the mill stream near Glastonbury.

It is likely that originally the Pillrow Cut was a natural waterway, which in medieval times drained the Meare Pool water into the Axe, and was not connected with the present lower Brue. Perhaps in the first half of the fourteenth century the Cut was straightened and canalized from Mark to Rooksbridge for both navigation and drainage, and in the early sixteenth the upper and lower Brue channels were connected. From the churchwardens' accounts of St John's Church, Glastonbury, for 1500, can be traced in detail the transport of new church seats from Bristol via Rooksbridge and the Cut to Glastonbury. There was, of course, also local traffic in corn, fish, wine and other needs of the abbey, not only on the Cut, but on a network of minor drainage and navigation waterways round Meare Pool. After the abbey was dissolved, the Cut was no longer maintained and slowly disappeared from the maps (part is now called the Mark Yeo), while Meare Pool also disappeared.

* There is another River Axe in south-eastern Devon, which flows past Axminster into the English Channel.

The Axe probably continued to be navigated to Rackley, and still more to Lower Weare on the Bridgwater–Axbridge road, and it is likely that in the eighteenth century coal and other goods were unloaded there to be carried to Axbridge. But when the French wars encouraged agricultural improvement, minds in Somerset turned to drainage and enclosure. In 1801 an Act set up Commissioners to drain the district connected with the Brue, and in the following year that round the Axe. Under this second Act, cuts were made to straighten the river, and in 1810 a large tide-sluice was built at Bleadon bridge, which blocked the river above it to all navigation.[1] When the Bristol & Exeter Railway's Act of 1836 was passed, it provided that, because the bridge to be built over the river at Batch would interfere with navigation on the $1\frac{1}{2}$ miles of river thence to Bleadon, the railway company should build a tide-sluice under the bridge, and wharves and approaches on each side of the river below it. Coal smacks from Wales were apparently still discharging their cargoes below the bridge at the turn of the century.[2] On the drainage waterways connected with the Brue, turf boats 17 ft long, about 5 ft wide, and carrying half a ton of turves, have worked to our own day: perhaps half a dozen survive.

Galton's and Brown's Canals

Two small canals once branched from the Brue. Galton's was built by Mr E. Galton[3] in the early nineteenth century under the Brue Drainage Act of 1801, from the river north-west of Meare village past Peacock Farm for $1\frac{3}{8}$ miles to the North Drain. From the river there was a short embanked section, then a small lock, the remains of which are still to be seen, and then the canal, some 16 ft wide, ran at the level of the ordinary rhynes of the area. Galton is said to have carried over 30,000 tons of silt, collected on the wide berm between the brink of the river and the floodbank, to the peat lands of the North Drain area. In 1830 the Commissioners of Sewers ordered him to repair floodbanks near the canal. After the middle of the century it seems to have been little used, and in 1897 the drainage authority built a wall across the end of the canal, with a tidal flap to allow water to discharge into the river. Part of it is now a drainage board rhyne.

About $1\frac{1}{4}$ miles downstream there is another canal, about a mile long, from the river to, or nearly to, the North Drain. It does not seem to have had an entrance lock, though there may have been a sliding sluice door. It was probably dug about the same

time, and for the same purpose, as Galton's Canal. It is now partially filled in.

Both canals were dug along the lines of older ditches made
either for drainage, or to divide fields under enclosure Acts.[4]

Glastonbury Canal

The drainage of the lands round the River Brue was under-
taken, as we have seen, in consequence of the Act of 1801. It
seems to have cost about £60,000, and largely to have been completed by 1807.

Nearly twenty years later, John Beauchamp jun., of West Pen-
nard, and Richard P. Prat, who with his brother Samuel was a
solicitor in Glastonbury, probably took the lead in suggesting
that a canal should be built from Highbridge to the town:[5] in
August 1825 a local newspaper reported the intention of bringing
in a bill, and observed that the plan would facilitate the inland
trade of the county, and reflected considerable credit on the gentlemen of Glastonbury.[6] Much discussion followed about the route.
In 1826 Richard Hamnett, a local surveyor, suggested alternative
routes. One, along the Brue, was rejected; for the other he proposed mainly to make existing drains navigable, with two 20-ft
broad locks on a canal section, a branch to near Street, and a float-
ing basin at Highbridge. A hopeful prospectus said this could be
done for £10,000, and estimated a profit of 10 per cent. There
were criticisms, so Beauchamp himself did a survey, and thought
it would be feasible to run a canal on the level from Highbridge to
Glastonbury, with two stop-locks. His estimate was £15,234.

The plan chosen by the promoters was nearer to Hamnett's
alternative. The navigation was to start at a basin by the mill
stream in Glastonbury, turn sharp north, cross the old channel of
the Brue by a lock down and another up again, and then proceed
by Cuckoo Ditch to Ashcott Corner. It would then follow the
South Brue Drain, with a lock at Shapwick, until it entered a new
canal to be built to the Brue at what was then called Cripp's House.
Thence it followed the river to Highbridge. There, the old winding channel had been by-passed under the 1801 Act by a new
straight cut and tide-sluice called the Brue Drainage, but this was
not to be used; instead, the old channel was to be reopened to
provide wharf space. Below, where the Brue entered the tideway,
a sea-lock would be built.[7]

At a meeting at the Town Hall in late February 1827, an engineer called E. T. Percy spoke in favour of the project, Samuel

Prat said that £14,000 had been subscribed towards Beauchamp's new estimate of £18,000, and it was decided to go ahead with a bill.[8] Subscriptions had risen to £15,250 by the time the bill reached Parliament; of this sum, all but £900 was contributed from within Somerset, most of it from Glastonbury itself. The Prats had put up £2,000, but in the name of Miss Ann Prat.[9] Newport coal shippers and those behind the Monmouthshire Canal supported the scheme, but not with cash, except for £350 from J. H. Moggridge, a committee-man of that company, and £100 from Sir Charles Morgan of Tredegar.

The Act[10] was obtained in May 1827. Power was given to raise £18,000 by shares or, rather unusually, by promissory notes or borrowing from the Exchequer Bill Loan Commissioners, and a further £5,000 on mortgage if necessary, the works to be completed within five years. Because the drainage of the Brue was in the hands of Commissioners of Sewers, the company were required to invest £1,000, to be at the disposal of the Commissioners in repairing or altering drainage works as a result of the building of the canal. They were suing for this money in 1831.

It soon became clear that the plan for which an Act had been obtained was impractical, and in 1828 Rennie was consulted upon what it was best to do. He considered three possible sizes of waterway: one 10 ft deep to take large sloops and small brigs of 120–140 tons, which he thought beyond Glastonbury's needs; one 8 ft deep, to take 40–60-ton coasters that worked in the Bristol Channel; or a 5-ft barge canal which would necessitate transhipment at Highbridge. He strongly recommended the second, at an estimated cost of £28,720, with locks 64 ft × 18 ft. This provided for a cut 44 ft wide at top. The two Brue locks were to be eliminated, and an 18-ft wide 3-arched aqueduct built instead, with a syphon for the river. There would be another single-arched syphon aqueduct over the South Drain, and a single lock at Shapwick.*

This plan was carried out, with two changes: that the canal above Shapwick was cut 10 ft deep to bring the surface of the water below land level, presumably leaving the size of craft the same, and the sea-lock was brought back to the upper end of the old channel, a little below its upper junction with the Brue Drainage. This change meant that Highbridge wharf was now below the lock, and therefore tidal; presumably the coasting trade had objected to paying a lock charge to reach Highbridge. The very sharp turn in the canal just short of Glastonbury basin, which Rennie had recommended should be eased, was built as originally

* About ¾ mile on the Highbridge side of the former Shapwick station.

planned. One assumes the necessary land could not be obtained, but the angle must always have been a difficulty to craft rounding it. Passing places were provided at intervals along the line.

The actual cost of construction must have been about £30,000, though the Act only authorized £23,000. The Brue aqueduct alone is said to have cost £3,000. Probably some of the money was never raised, and the debts never paid: certainly Rennie, who had not only done the final survey, but provided specifications for locks, aqueducts and a bridge, was in 1837 still owed £367 out of his bill of £567, in spite of having frequently asked for it. He was then threatening legal action, but the proprietors must have been used to that.

The canal was opened on 15 August 1833.*

Early in the morning, the road leading from the town to the basin, was thronged with spectators, and the whole population appeared to be on the *qui vive*. About seven o'clock, the Company's barge, the *Goodland*, and the beautiful yacht, the *Water Witch*, filled with the most respectable persons of the neighbourhood, set sail for Highbridge, accompanied by a band of music, the bells ringing a merry peal; the party returned about six in the evening. On Monday last, as early as five o'clock in the morning, the Company's barge, together with a new yacht, the *St. Vincent*, (the property of J. Vincent, Esq.) having on board nearly 150 persons, principally of the respectable trading classes, took another trip to Highbridge, and returned in the evening, amidst the ringing of bells, firing of cannon, etc. We understand it is the intention of a large part of the gentlemen of Glastonbury, shortly to give a public dinner in a commodious booth, to be erected near the aqueduct in the vicinity of the town.[11]

It started well, and a year later there was talk of extending it by a railway to Yeovil, Sherborne and Dorset. Phelps, the historian of Somerset, writing soon afterwards, gives the following description of the canal in use:

A direct communication has been opened between this town and Bristol, Gloucester, Newport, Caerdiff, and other ports in the Bristol Channel. By this canal, the manufactures of Manchester and Birmingham, pottery, salt and other articles, shipped at Gloucester, coals and iron from Newport and Caerdiff, can be brought direct to Glastonbury; and in return, the fine elm timber of the district, paving stones, corn, cheese, cider, etc.,

* A print of the opening of the canal is reproduced facing p. 49 of the 3rd (1966) edition of my book, *British Canals*.

VII. Parrett Navigation: (*above*) Below Langport in 1938. The navigation lock is on the right, and floodgates on the left; (*below*) basin and warehouse at the terminus of the Westport Canal in 1967

VIII. Grand Western Canal about 1951: (*above*) The lift cottage at Trefusis; (*below*) Halberton wharf

can be transferred to the populous and manufacturing districts of South Wales.

Another important advantage to the marsh country, is that the flood waters now pass off rapidly through the lock-gates of the canal, which are opened when the water is high, and a double outlet is made in addition to the original sluices; thereby relieving the marsh of its waters, in about as many days as it required weeks heretofore.

In January 1840 the Prats' firm failed, with serious effects on the town's prosperity, their difficulties being attributed partly to their investment in the Glastonbury Canal.

In 1842 the Bristol & Exeter Railway was opened to Highbridge, and perhaps brought some additional revenue from transhipment business. Then, at the end of 1846, the secretary of the railway company was sent to Glastonbury to treat for the canal's purchase 'if it is to be obtained at the price specified to him', apparently because the railway feared 'a narrow gauge Company looked to the site of the Canal in this district as opening up a probability of securing a footing here';[12] presumably this was the South Western. In June 1847 the company decided to buy, and in 1848 agreement was reached at a price of £7,000, and an authorizing Act passed. Payment was made in 1850, the canal company being dissolved in 1851.

The railway company's shareholders' meeting in early May 1848 was less than enthusiastic. The chairman explained that the canal was almost the only water communication with their main line, implied that others wanted to get hold of it, admitted that its current revenue was only £300 pa, asserted that there would be a net income to the company, even though there would be some liabilities in addition to the purchase money, and hoped that the railway would be able to increase the water traffic. Some in the hall thought the purchase unjustified. One said: 'Had anyone ever seen any boats or traffic on the canal? He had been there time after time, and had never seen a boat in his life.' Another said that some time ago a mortgage of £7,000 or £8,000 on the canal had been sold for £3,000 or £4,000 (this may have been when the Prats failed), and therefore it was monstrous to pay £7,000 for it.

The Act[13] required the canal to be kept open and in good repair, after the preamble had said that 'it might be more economically and beneficially worked in connexion with the Railways and Works' of the Bristol & Exeter. For some time past, however, the peat bed of the canal had been rising as it became waterlogged, while lack of dredging had caused mud to accumulate. This made

F

it shallow and hindered traffic. The gates at Highbridge were therefore kept closed to hold back the fresh water and give greater depth. But in 1850 this caused low-lying lands in the vale to become flooded, the dyke reeves at Chilton Polden to present the canal as filled with mud, and the Commissioners of Sewers as the drainage authority to order the gates at Highbridge to be opened until tide and current had thoroughly cleansed the bed of the waterway. This in turn made it more difficult to navigate.

The railway company protested that nearly everyone from the canal company who had signed the agreement to sell was also a Commissioner of Sewers, and that the Commission's act in opening the Highbridge gates would take away almost all the value the canal had for them.[14] This situation must have encouraged the Bristol & Exeter to make an agreement of the same year to resell the canal to their creation, the newly-promoted Somerset Central Railway, for £8,000 of its shares and a promise to subscribe £2,000 more, with the intention of abandoning it. The enabling Act[15] of 1852 enacted that within three months of its passing, the Bristol & Exeter were to transfer the canal, and the Somerset Central were to accept it. Power was given to abandon it, and use any part for the new line, except for the portion west of the Bridgwater–Bristol road at Highbridge. This and the harbour were to be maintained by the railway under the jurisdiction of the Somerset Commission of Sewers. A short diversion of a part of this retained section was authorized, a new channel to be made and the line built over the old one.

The transfer took place on 16 September 1852. The new railway's broad-gauge line was then built from Highbridge along the canal bank for most of the way to Glastonbury. Evidently the waterway carried construction material, for in August 1853 the Somerset Central reported that canal traffic was greater than ever before. Gross receipts for the year ending 30 September 1853 were £312, and maintenance £182, but whether the railway paid for its own material to be carried I do not know—probably not. The navigation between Glastonbury and Shapwick bridge was closed on 1 July 1854. As the railway opened on 17 August, it is likely that a gap was left in the embankment at a point where it crossed the canal near Ashcott, until the last minute. Probably the rest of the abandoned portion became disused soon afterwards.

The Somerset Central, but not the canal, was leased to the Bristol & Exeter for seven years from the opening date, and later became part of the standard-gauge Somerset & Dorset. The portion at Highbridge reverted to the Commissioners of Sewers, and was abandoned in 1936.

River Parrett
Ivelchester (Ilchester) & Langport Navigation, and Westport Canal

Ships have always come up the River Parrett to the port of Bridgwater, and for a long time craft passed beyond the town, either continuing up the Parrett to Langport and beyond, or turning off at Burrow bridge to ascend the Tone towards Taunton.

The Parrett is tidal nearly to Langport, and has a small bore which used to be of some help to barges. The river was being well used in Charles I's reign, and coal and culm were being imported. The firm of Stuckey & Bagehot—from which was later to spring the well-known Stuckey's Bank—was formed in 1707, and thereafter increasingly traded on the river, on the sea from Bridgwater,[16] and up the Severn.

There was also navigation beyond Langport, farther up the Parrett to Thorney, and up the River Yeo or Ivel. There is evidence of two Roman wharves at Ilchester,[17] but in modern times it is likely that boats only reached Ilchester wharf—on the left bank, below the bridge—at times of flood. At other times they unloaded at Pill bridge whence a lane led straight to Ilchester, or Load bridge near Long Sutton, whence coal and other goods were carried to Ilchester, Yeovil, and the district round. Mention is made of Pill bridge by Thomas Gerard in his *Particular Description of the County of Somerset* (1633): 'From Ivelchester the river passeth under Pillbridge, whither are brought up boates and crayes from Langport and Bridgewater.' Langport bridge was, however, a complete obstacle, and all goods had to be transhipped across the bridge from the larger barges working down to Bridgwater to smaller craft carrying some 7 tons that used the upper rivers.

When the Dorset & Somerset Canal to link Poole and Bristol was being discussed, it was suggested in February 1793 that it might be better built from Poole by way of Wareham, Dorchester, Yeovil and Ilchester to join the proposed Bristol & Taunton Canal below Langport, whence there would be access not only to Bristol, but to Taunton and the Grand Western also.[18] Such a scheme would, of course, also link the navigations above and below Langport bridge. This idea developel in 1794 into a possible navigation from the Parrett below Langport to Ilchester, possibly to Yeovil, possibly still to the Dorset & Somerset. William Bennet, who had surveyed the Dorset & Somerset and was shortly to become its engineer, examined the route, with Charles H. Masters as his surveyor, estimated it at £5,102, and told a meeting in October,[19] with the Earl of Ilchester in the chair, that landowners need not worry lest their property should be injured, presumably because the projected locks were to have very small rises.

In the following year an Act[20] was obtained authorizing a line 8¼ miles long from the Parrett just below Langport to Ilchester, which would by-pass the obstructive bridge by using the Portlake rhyne through Little Bow bridge (to be rebuilt to take barges), in the town, whence a new cut would be made for ¾ mile to Bicknell's bridge on the Yeo, a little above its junction with the Parrett, which would be dredged and improved to Ilchester. Seven locks were planned, each 75 ft × 7 ft 6 in, one on the Portlake rhyne below Little Bow bridge, two on the cut from the rhyne to Bicknell's bridge, and four on the river up to Ilchester, all seven achieving a total fall of only 16½ ft.* The Ivelchester & Langport Navigation Company were authorized to raise £6,000, and a further £2,000 if necessary.

Work began at once, with Josiah Easton, a local man of strong opinions, in charge. By April 1797 the company had called the whole £6,000. They had straightened and deepened the Portlake rhyne, rebuilt Little Bow bridge (with a contribution of £87 from Langport corporation), and made part of the cut to Bicknell's bridge. If any locks were built, however, no traces now survive. At the end of 1796 Whitworth had been called in to go over what had been done, and estimate the cost of finishing the work. Given the time's rising prices and scarcity of money, the sum must have been more than the shareholders' meeting of February 1797 cared to face, for they made a final call to pay their debts, and left the work unfinished and useless until easier times. It was never restarted, and the company soon ceased to meet.[21] About this time, in 1800, a more extensive scheme was prepared for making a canal and drainage works from Dunball a little below Bridgwater on the Parrett to Yeovil, via Somerton and Ashington. It came to nothing.

There was now a considerable traffic of some 50,000 tons a year over the 12 miles of tidal river between Bridgwater and Langport in 15–20 ton barges, but for an average of four days in a fortnight it was stopped by neap tides or floods. Stuckey & Bagehot had

---

* What is probably an earlier plan proposed three (possibly four) half-locks and one (possibly two) locks on the river section, instead of the four locks that were advertised for tender.

almost two-thirds of the Langport trade, and two other local merchants the rest. There was also traffic beyond Langport, mainly to Thorney on the Parrett, where two merchants traded, or to Load bridge on the Yeo, where another man had a wharf. But the old nine-arched Langport bridge was a complete obstruction to any navigation through it. In the early part of the nineteenth century it seems that Henry Lovibond built a railway to carry barges past the bridge at a cost of £400. However it worked, it cannot have done so successfully, because at the time of the Parrett Navigation bill goods were transhipped 'on Men's shoulders, and carried on Planks under the Arches'. Some got damaged, and bulky goods could not be handled at all. Above the bridge, they were loaded into 5–7½ ton craft, usually worked in pairs, to go farther up the rivers. This upper river trade included corn and manufactures from Yeovil, and goods from Bridport for the American market. Langport at this time was prosperous, and the population had risen to its highest point in 1831.

| 1801 | 754 | 1821 | 1,004 |
| 1811 | 861 | 1831 | 1,245 |

Hanning had, as we have seen, first suggested a canal over Westmoor in 1831 as part of an enclosure scheme. When the Chard Canal promoters met him and some Langport men in 1833, they understood that the Westmoor proposal had been dropped. Certainly there was no opposition from Langport to the Chard bill of 1834. Soon afterwards, however, a group, including a Stuckey and some Broadmeads, thinking that the canal would draw off trade from the Parrett, began first to consider improvements to the navigation of the river above Langport, and the building of a canal over Westmoor towards Ilminster and Chard, and later also of improving the Parrett below Langport.

In 1836 they and their supporters sought an Act to improve the navigation of the river from Burrow bridge past Langport to Thorney, and from Muchelney to make the River Isle navigable for about a mile and then to cut the Westport Canal to end where the Barrington road left the Langport to Ilminster turnpike, where a basin and five wharves were later built. In this way it was hoped to save from the Chard Canal a good deal of the Parrett's traffic arising round Ilminster.

The Act[22] established the Parrett Navigation Company, authorized to raise £10,500 by shares and a further £3,300 by mortgage. But it was only obtained after bitter opposition from the Chard company, who thought a promise made to them had been broken, helped by their allies the Bridgwater & Taunton and those concerned with the Tone trade; it cost the Parrett promoters £4,212 to get. At the first meeting on 28 July 1836, the clerk was instructed to write to the Chard company 'to enquire whether that Company will be disposed to form a junction between the end of this Canal at Park Gate and their Canal at any point between Ilton and Ilminster'. At this time also it was suggested that the Yeo should be surveyed to Ilchester as an encouragement for others to improve that navigation, but nothing came of either initiative.

William Gravatt, at that time assistant to Brunel on the building of the Bristol & Exeter Railway, was made engineer, with Charles Hodgkinson as his resident, and Henry Draper, an auctioneer, was made superintendent with the curious proviso that if he were away from work because of his auctioneer's business he must find a substitute. It was intended to build locks at Stanmoor above Burrow Bridge, Langport, and Muchelney at the entrance of the Isle, but a fourth, at Oath below Langport, was soon added. There was also a half-lock at Thorney. Part of the unfinished cut of the Ivelchester & Langport company was used as a flood drain. The Chard Canal was of course being built at this time, and it was 'Desirable to finish the works with all possible expedition', but money was running out, and a second Act had to be obtained in 1839 to authorize the raising of another £20,000, and an increase in the tolls.

The passing of this Act produced an odd incident, the forgery of a petition against it. This forgery, which apparently consisted of cutting off the list of signatories from one petition and attaching it to a stronger one, was investigated by a Select Committee of the House of Commons, which reported[23] that from the minutes of evidence taken before the Committee it appeared that Benjamin Lovibond was sent to parliamentary agents in London with a view to obtaining a suitable petition against the Bill, as the merchants wished to oppose any increase in the tolls. He sent these latter a draft petition which they altered and returned to him duly signed. The alterations did not suit his own interests and he substituted the original wording, attached the signatures, and presented the petition.

The navigation from Stanmoor to Langport was fully opened on 28 October 1839. The company now began to pull down and rebuild Langport bridge in order to improve the upper river. When the contractor looked like getting behindhand, it was minuted that there would be

... serious injury to the Company if it be not finished by that time,* as well in the costs of their superintendent of the works, as by injuring them in their competition with a rival company and otherwise.[24]

The navigation above Langport and the Westport Canal were completed by about the end of August 1840, and the bridge on time, nearly a year ahead of the Chard Canal. The work had altogether cost £38,876, the excess over the authorized total of £33,800 being covered by certain receipts and by incurring debt. The cost of rebuilding the bridge† was paid for as to £500 by the Langport corporation, and for the rest by a loan repaid by a special bridge toll which was extinguished in January 1843. The 210 original shares were of £50 each, but those authorized under the second Act, though nominally of £50, were issued at £10 only. The company had therefore a nominal capital of £10,500 in old shares and £85,000 in new shares, though only £27,500 had been subscribed as share capital. When dividends were paid later, they were declared on the total nominal value of the shares.

Because the Chard Canal was not yet open, the takings in tolls for the first financial year August 1840 to July 1841 were £2,175 1s 4d. At the end of May 1842, however, that canal was ready, and in August the Committee had to report that

The joint operation of the Bridge toll and the Chard Canal has been to seriously injure our traffic—the latter must for the present at least be taken to be a permanent injury. . . .[25]

The tolls for the next year were £1,703, and the average for the following five years £1,508. The company paid no dividends in its early years, but set itself to repay debt, except the share capital; this it did by 1852, a year before the competing railway was opened.

There are few records of the tonnage carried, except a note that for the year ending 3 July 1839, just before the navigation was fully open, the figure was 55,190 tons. An estimate of future traffic then showed 15,000 tons to Langport, 35,000 tons to the Westport Canal, 5,000 tons to Thorney, and 15,000 to Yeovil, these being probably landed at Load bridge. The largest barge carried 24 tons, but most were a good deal smaller than this. For the calendar year 1843 the following figures show the tonnages carried by Stuckey & Bagehot boats:

* 1 March 1841.
† I think £4,249, including £260 compensation for removing the tramway.

	Tons
Langport	27,605¼
Load bridge	9,302¼
Thorney	324
Westport	10,402

Stuckey & Bagehot contributed about two-thirds of the tolls, so if we increase these figures in the same proportion, we get about 70,000 tons a year for the whole navigation. The Westport figures, when compared with the earlier estimate, show the effect of the competition of the Chard Canal in the supply of the country towards Ilminster.

When the navigation had been completed, it seemed good to some of the promoters, the Broadmeads, who held the majority of the 1839 shares, to think once more of making the River Yeo navigable to Ilchester. The old Ivelchester & Langport company was no longer in existence, but Nicholas Broadmead got the list of shareholders from the son of the late clerk and wrote to them offering £5 each for their £50 shares. Many were dead, some did not reply, some smelt a rat and refused to sell, and 29 shares passed into his hands, with the prospect of 38 more out of the total of 120. He thought it possible to make the river navigable now for £2,000, apparently without locks, because the work that had been done on the Parrett, and especially the building of Langport lock, had already improved it by increasing the depth of water up to Bicknell's bridge. He estimated that it might carry 14,133 tons a year and earn a revenue of £500, this to include, of course, what was now going free of toll from the junction with the Parrett Navigation to Load bridge. His idea was that the 1839 Parrett shareholders should gain control and work the Yeo themselves. He suggested a method of forcing the remaining shares out of their original owners' hands in a not very ethical letter to Philip Broadmead:

> . . . there is power in the old Act to raise a further £2000 by calls on the shares, and if the shareholders do not pay their quotas, their shares will be forfeited. If you were to call for the £2000 (without telling the shareholders what our plans are) almost all of the holders . . . would forfeit their shares or accept £5 per share rather than go on. . . .

A meeting of Ivelchester & Langport shareholders was called at Ilchester by the son of the late clerk in conjunction with the Broadmeads to restart the company and appoint a clerk, but there is no evidence of what happened at it. The absence of any further mention of the Yeo in the Parrett records makes it likely that the

Broadmeads in the end did not attempt to take tolls or to spend money on the river.

In August 1845 there is the first sound of a railway threat to what had promised to be a moderately prosperous navigation. The committee report to the annual general meeting that they have opposed the bill for the railway from Yeovil through Langport to join the Bristol & Exeter Railway at Durston between Bridgwater and Taunton. They go on:

> Danger has been apprehended of your losing the Yeovil Trade in consequence of the Railway and a question is raised whether the Coals brought to the Port of Bridgwater will be carried wholly by the Railway thence to Yeovil, Sherborne and the interior of Dorsetshire or whether (your tolls being remarkably low for the Yeovil trade) you may not secure the carriage on the river as high up as the point where the railway crosses it between Muchelney and Westover Farm. In order to obtain this advantage it would be necessary to purchase land and make a Station near the place of crossing. . . .[26]

A few months later the committee were willing to sell the navigation to the abortive Bath & Exeter Junction Railway rateably to the price that concern was offering for the Somersetshire Coal Canal.

In 1847 the Yeovil and Durston line was begun. The railway mania was over, money was short, and in 1848 there was a falling off of tolls

> occasioned chiefly by the stoppage of the works on the Railways near Yeovil which has induced the Owners of Carts and Horses lately employed thereon to bring down Coals from Mendip at an unusually cheap rate.[27]

Railway building began again in 1852. When the line is completed

> your Committee cannot but fear that your receipts will be considerably diminished by the competition. . . . Your Committee cannot report any arrangement whatever with the Railway Company either as to making a communication between the River and the line or otherwise and your Committee believe it to be the determination of the Railway Company not to allow of any such Communication.[28]

The line was opened on 2 October 1853 for passengers, and at the end of November for goods. In August 1854 the Committee reported happily that a mutual understanding existed with the railway company that

> the whole or a greater portion of the heavy merchandize brought to Bridgwater by Ship and intended for Langport and

the Districts beyond is still brought to Langport by water and there put on the Railway, and that a communication between the Railway Station and the River has been formed.[29]

A year later there was less happiness, for

... your Committee have to add that such heavy Merchandize has been diminished by reason of all the places on the branch line being now supplied with Coal from the Radstock pits which is conveyed from thence by the Railway at so low a charge, as to enter into serious competition with the Welsh coal.[30]

The effect of railway competition can be seen in the record of tolls:

Year July–June	Toll Receipts £
1848–9	1,485
1849–50	1,421
1850–1	1,429
1851–2	1,469
1852–3	1,686
1853–4	1,440
1854–5	920
1855–6	832
1856–7	868
1857–8	673

There was a brief revival to an average of £759 for the four years ending June 1862, after the Westport Canal had been thoroughly dredged, and then a slide downwards to £347 for the year ending June 1871. In this year a committee of landowners concerned with drainage produced a report which complained of the state of the river and its drains, upon which the committee minuted:

... The Company will endeavour to rectify the matters complained of, but the Navigation Committee must remind the Drainage Committee that the Funds at their disposal are very limited. . . .[31]

Drainage considerations were now more important than those of navigation. The end came on 19 February 1875, when, Westmoor being partially flooded, Mr Thomas Mead opened Langport lock and forbade the lock-keeper to close it. This was done with the concurrence of the local landowners and occupiers, because of the Navigation Company's failure to repair the drainage culvert under the river at Huish bridge. The act was legal in such an emergency; the company had not the money to repair the culvert, and therefore they ceased to levy tolls. Two years later, in 1877, the locks were still open, and several summonses had been issued against the navigation by the Commissioners of Sewers for failure to maintain drains. This being so, it was decided by those promoting the Somersetshire Drainage bill of that year, which reorganized the drainage authorities of the county, to include in it power to take over the navigation. The Act was passed in the same year, and from 1 July 1878 the navigation vested in the new Commissioners without payment. The relevant clause made it lawful for the Commissioners to abandon any or all of the navigation, but they took it in a wider meaning, for when in December 1880 a petition was received from the inhabitants of Barrington and the villages round, asking the Commissioners to keep open the Westport Canal, they replied: 'That inasmuch as the Commissioners have no power under their Act to keep open the Westport Canal for trade purposes, they were unable to accede to the petitioners' request.'[32] The river was still used to some extent, even though it had reverted to its state before 1836. Boats navigated to Langport and even occasionally to Ilchester past the turn of the century, and in 1911 there is a reference to barges coming up to the Aller footbridge near Oath (below Langport) on the river's bore.[33]

The Navigation Company paid no dividend till it had repaid its debt. Then, for the year ending June 1853, a first dividend of 14s per £50 share, or nearly 5 per cent on the money actually subscribed, was paid. The railway opened during the next financial year, and the dividend was halved; thenceforward it fell away to a last payment of 1s per £50 share for the year ending June 1872. The company had done better than its old rival the Chard Canal: it had paid dividends, and it had lasted ten years longer.

Before we leave the Parrett it is interesting to note that throughout the lifetime of the navigation the old firm of Stuckey & Bagehot were the largest carriers, and their boats, later run as Bagehot & Co., and after that as the Somerset Trading Co., remained to the end.

Dorset & Somerset Canal

Farther east than the others, a bigger canal had been promoted during the mania and then, while under construction, had been caught by rising costs and scarcity of capital. The Dorset & Somerset or Poole Canal had been projected in 1792 to link Poole and Bristol by way of Wareham, Sturminster Newton, Wincanton, Frome and the River Avon at Bath, with a branch to the Mendip collieries round Nettlebridge, the main purposes being to carry coal southwards to Dorset, Wareham clay northwards, and agricultural produce generally. A meeting at Wincanton in January 1793 opened a subscription list.[34]

Controversy then began between groups of supporters and landowners upon the best route for the section south of Stalbridge, whether by Blandford and Wimborne, or by Bagber,Mappowder, Plush, Piddletrenthide and Wareham. Whitworth was called in to make a survey. He reported in September[35] that a line 37 miles long from the intended Kennet & Avon Canal at Freshford near Limpley Stoke to Stalbridge would cost £100,000 and be well supplied with water. Thence by Blandford would be 33 miles for £83,000, or by Wareham 30 miles for £91,000, but the latter route would have water difficulties. Saying how busy he was, Whitworth then stepped out of further work for the projectors, and recommended William Bennet of Frome, who had surveyed for him, as engineer. The projectors, realizing that their main traffic would be coal from the colliery branch, then started to canvass mine and land owners.

It was not until mid-1795 that Bennet finished his detailed survey of the main line and the branch from Frome to the Nettlebridge collieries. By now the Somersetshire Coal Canal, also joining the Kennet & Avon at Limpley Stoke, was being built, and was indeed mentioned as a source of coal supply in the notice for a promoters' meeting in July;[36] it seems odd that they still thought a separate and expensive colliery branch of their own necessary. The meeting approved almost all Bennet's line, and his estimate of £200,000, though later the proposed junction with the Kennet & Avon was altered to Widbrook near Bradford on Avon, nearer that canal's junction with the Wilts & Berks. Two short branches were now added, to Wareham and Hamworthy near Poole.

The northern part of the line, with its prospective connections to the Kennet & Avon, Wilts & Berks, and Somersetshire Coal Canal, looked promising, but the southern end now ran into serious landowner opposition. It may have been because they wished to avoid this, and possibly also because they had doubts about raising so much money, that the promoters in the end sought a bill for their canal only as far south as Gains Cross below Shillingstone, south-east of Sturminster Newton, where it was to end beside the main road to Blandford. They probably thought that coal could well be distributed thence by road. Bennet estimated the shorter line at £146,018, and an Act[37] was obtained in 1796 for a main line of 49 miles, and an 11-mile branch, which gave the company a capital of £150,000 and power to raise £75,000 more. But

the spirit of speculation was so completely changed by the altered condition of the country, the influence of the French Revolution, the large sums raised by Government on loan, and the high prices which accompanied these events, that not more than £80,000 was subscribed, and out of that sum, only about £58,000* was actually received by the Company.[38]

In fact, £79,200 had been subscribed, the biggest shareholders being Richard Perkins (£8,500), John Lambert (£4,000), the Rev. Samuel Farewell (who had taken the chair at many of the promotion meetings) (£4,000), Job White (£3,500) and John Billingsley (£3,500).

The proprietors decided to build the branch from the collieries to Frome,† estimated at £30,584, which under their Act they were bound to complete first, reckoning that the money in hand would suffice. Nearly 8 miles were actually built from the bottom of Stratton common about half a mile west of Edford through Coleford and Vobster towards Frome, including a three-arched aqueduct at Coleford, a smaller one at Vallis Vale, and an unfinished tunnel at Goodeaves Farm. Some sections were filled with water, and on that nearest Nettlebridge a barge, probably one of the contractor's, was launched.

The main line, and also the branch, involved heavy lockage: 264 ft on the latter. James Fussell, with his ironworking knowledge, interested himself in the possibility of using vertical lifts instead of locks. He probably knew that John Duncombe, an engineer on the Ellesmere Canal, had about 1790 invented a counterbalanced lift, and that Messrs Rowland and Pickering of Ruabon had in 1794 taken out a patent for one of float type.[39] Nearer home, Robert Weldon was proposing one of yet another pattern for the Somersetshire Coal Canal at Combe Hay.[40]

But he must have been experimenting much earlier than that, for the documents accompanying the bill provide for 'caissons' to be used on the branch, and together with locks on the main line. The word suggests that Weldon's type was then in mind. At the top of Barrow hill, Fussell had the first lift built so that it could be tested. It had a lift of 20 ft, and took 10-ton boats. In principle, it was similar to those James Green later built on the Grand Western

* This seems to be under the true figure, which the clerk stated in 1803 to be £65,000.
† It should be remembered that Frome was then at the height of its prosperity, and had an estimated population of 8,100 in 1790 against 5,800 for Taunton in 1801.

Canal (see p. 104), with two counterbalanced caissons into which boats could be floated. The main differences were that the suspending chains, instead of being attached to a bar above each caisson, were run under wheels mounted on the caisson and then fixed to the upper part of the lift structure; that wheels placed on the sides of the caissons ran on vertical guide rails to steady them as they rose or fell; and that the water used to overbalance the uppermost caisson was not run into the caisson itself, but into a chamber beneath it.[41]

Tests of the Barrow Hill (or Mells) lift were made in September and October 1800. At that on 13 October, the local reporter said:

It answered the design perfectly to the satisfaction of a great number of spectators: among them were many men of science, impartial and unprejudiced, who after its repeated operations and those without the least difficulty or mischance, and inspecting minutely every part of the machine, were unanimous in declaring it to be the simplest and best of all methods yet discovered for conveying boats from the different levels and for public utility.[42]

In January 1801 the company announced that they would build five more on the same section, and called for tenders. Pits for four farther down Barrow hill were excavated, and some masons' work done. But by the following year all their money had been spent, and about 1¾ miles of the branch remained to be cut, to join existing portions and complete the line to Frome, as well as the unfinished lifts. They called in an engineer called Underhill (perhaps William Underhill of the Dudley Canal) who gave a report to a shareholders' meeting in March on the expenditure needed to complete the branch to Frome, and upon the present and likely future supplies from the local collieries. It seems to have encouraged them[43] to get a second Act in 1803 to enable money to be raised by promissory notes with the option of conversion into shares, to pay off debts of £1,100, and to finish their line, if necessary by building a railway on any part of it to lessen the cost, but the money was not forthcoming, and the works were abandoned, though the company was still alive at the end of 1825, when it was thinking of building a railway to carry coal into Dorset.[44] Much of the canal line can still be traced.[45]

CHAPTER VIII

The Grand Western Canal

ENTHUSIASM for canals reached Devonshire in 1792, and many proposals were made. The most important of these was the Grand Western from the Exe to Taunton to connect with the Tone Navigation or a projected Bristol to Taunton canal.

The first promotion meeting was held at the Half Moon Inn, Cullompton, on 1 October 1792, and was reported in the *Exeter Flying Post*:

Resolved unanimously: That the plan for making a navigable canal from Taunton to Topsham with a branch to Tiverton is clearly practicable at a very moderate expense, and that it will be productive of the very greatest advantage to trade, commerce and agriculture.

That among the advantages that may arise from making this canal, appears to be the easy communication betwixt the Bristol and the British Channel, instead of sailing round the Land's End, which requires various winds, and both in winter and war is a tedious and dangerous navigation.

That this canal will afford a safe, easy and expeditious conveyance for coals, lime, timber (and other materials for building), iron, cheese, salt, groceries, hardware, wool, etc.

That it is an object of national importance by the ready conveyance of timber to His Majesty's dockyards from the north of Devon, the counties of Somerset, Gloucester, and (particularly the Forest of Dean) Hereford, Worcester, etc.

That the general communication between Ireland and Scotland with the western part of England and the Bristol Channel would also be rendered much more easy and safe.[1]

The proposed line ran from Topsham on the estuary of the Exe up the Clyst valley, with a short branch to the Exeter road near Sowton; then by Clyst Hydon into the valley of the Culm. It passed Cullompton, to which there was a short branch, and ran a little south of Sampford Peverell, whence a fairly straight branch went to Tiverton. A feeder then came in from Culmstock and two reservoirs beyond, and just beyond Burlescombe another from two more reservoirs to the northwards. The main canal then passed by Runnington north of Wellington to Bradford and the river at Taunton about half a mile above the Tone bridge. The estimated cost was £166,724, including £22,229 for the Tiverton and Cullompton branches.

This first survey had been made by Robert Whitworth. John Longbotham, a pupil of John Smeaton who had been concerned with the first planning of the Leeds & Liverpool Canal, was then called in to make his own recommendations, and Robert Mylne to revise what Longbotham had proposed. At this point 88 of the 90 original subscribers were sufficiently optimistic about the project to double their shareholdings, so producing another £71,200. Finally, William Jessop, the most eminent canal engineer of the day, was called in to decide between Whitworth's and Longbotham's differing proposals for part of the line, to say whether further improvements could be made, recommend the right dimensions, report where it should begin and end, and whether it was likely to get enough business to justify the outlay. He in turn seems to have employed Hugh Henshall, (he who completed the Trent & Mersey and Chesterfield Canals after the death of his brother-in-law James Brindley), who used Mylne's earlier work.

Jessop's report was made to a meeting once more at the Half Moon Inn, Cullompton, on 28 November 1793, with Sir George Yonge in the chair. On his first assignment, Jessop chose Longbotham's line as shorter and cheaper, but was unenthusiastic about the Tiverton and Cullompton branches, which he suggested should be left out of the bill. On dimensions, one of his predecessors, probably Whitworth, had suggested narrow locks, but Jessop thought that, as the principal object of the waterway was to join the Channels, the company should provide for decked barges 14 ft 6 in or 15 ft wide, to carry 50 tons, which could go down the Exe to Exmouth or across the Bristol Channel to bring coal from Wales. The waterway should be 42 ft wide, except in deep cuttings or on embankments, and 5 ft deep.

On the last of his responsibilities he was interesting. 'To make this Canal productive to the Undertakers,' he said, 'it is essential that Coal should be obtained and conveyed at such a Price as to supply some part of the Consumption at Exeter: this will depend on . . . whether the Coals from Wales may . . . be charged with the Sea-Duty.' At that time coal from Cardiff and Welsh ports west of it paid the coastwise coal duty, but that from Newport did not.[2] Therefore it would be sensible to extend the canal beyond Taunton to Uphill, where the Newport coal could come, 'the easiest Line for a Canal that I have ever seen,' or from Nailsea, 'supposing the Bristol Canal in that line to be executed'. Then coal could be brought a good deal cheaper than by land carriage, and sufficiently below the cost by sea to obtain some part of the Exeter market, especially in small coal for burning lime and bricks. There were also markets for coal in the towns on the line, and at the great limeworks at Canonsleigh near Burlescombe, then selling 20,000 tons of lime a year as a necessary soil conditioner. He estimated a revenue from coal and lime of £8,311, and added:

In similar projects I have generally found, that if on Information fairly and accurately obtained, there appears on the Outset of the Concern that there will be Articles conveyed that will ensure Three per cent on the Expence, the progressive Increase of Carriage, which is always a Consequence of a Reduction in the Price of it, makes it a profitable Undertaking to the Adventurers.

But he added that success greatly depended on the extension from Taunton towards Bristol, and suggested that both lines might well form part of the same concern.

In December 1793 the committee approved the plan Jessop had recommended. But during 1794 they called in yet another engineer, John Rennie, who with two assistants made considerable changes in Jessop and Longbotham's line, and reported on the proposed reservoirs. Their project was now being opposed by Exeter corporation, who feared competition with the Exeter Canal, especially in coal, the loss of their petty customs receipts, and of landing dues and charges for the use of their crane at Topsham. They also complained that water from the Culm would prejudice the supply to the city and industries beside the Exe. The company assured the corporation that the Culm would not be affected, but had to give up the Sowton branch and agree instead to carry coal through to Topsham so that it could go up the Exeter Canal, a more expensive route which lessened their prospects of success.[3]

Early in 1796 they were willing to guarantee the corporation against any loss of dues, and, with a final estimate of £211,875 from Rennie they at last obtained their Act[4] in its final form, for a line 36½ miles long, with a rise of 326 ft 4 in. from the Exe to the summit between Burlescombe and Greenham, and a fall of 265 ft 1 in to Taunton, where 500 yards of the Tone were put under the company's control. There were after all to be branches to Tiver-

G

ton, 7¾ miles with a 78 ft fall, and to Cullompton, 2 miles with a 30 ft fall, and reservoirs at Lowdwells and Hemyock. The share capital was £220,000, with power to raise a further £110,000 if necessary.

But now the increasing tension of the war, the rising prices of labour and commodities, and the collapse of the Bristol & Taunton project to continue their line, and with it the inter-Channels scheme to connect with it, which would have brought additional trade, caused the plan to be shelved for ten years. Then, at much the same time, the Grand Western and the first Chard plan revived again, as confidence gained upon discretion. It was in September 1807 that the Kennet & Avon company were approached for support in getting the Grand Western started, but they were too financially involved themselves to help, and refused a similar appeal again in June 1808. In July of that year a meeting of Grand Western shareholders asked two of their number to investigate the possibility of beginning construction, and to prepare a report. Meanwhile many of the original subscribers had died or withdrawn, and 903 of the 2,200 original shares remained unappropriated. Early in 1810 certain proprietors of the Kennet & Avon agreed to do what the company had refused, and took these over, a clear indication that a through canal from London to Exeter was envisaged. In the following year a pamphlet[5] observed

> that it is proposed to make the Canal from Bristol to Taunton of the same dimensions as the Kennet & Avon and the Grand Western Canal: so that barges of fifty tons burthen may be laden at Exeter, and proceed to London (a distance upwards of 200 miles) without shifting their cargoes.

A general meeting in April 1810 authorized the beginning of the work. It seems to have been realized that the original capital of £220,000 would not be enough, for one of the subjects discussed was: 'To consider if £220,000 be enough, and to raise another £110,000 if necessary.' On 19 April 1810, the following news item in the *Exeter Flying Post* marked the beginning of an attempt to link the two Channels.

> Monday last . . . the great work of the Grand Western Canal was commenced on the summit level in the parish of Holcombe, on land belonging to Peter Bluett Esq., for which occasion the first turf was cut with all due form and ceremony by the Rt. Hon. Sir George Yonge, Bart., chairman of the meeting, assisted by the lady of John Brown, of Canonsleigh, Esq., who attended for that purpose, in the presence of a numerous body of spectators of all ranks, who testified their joy at the com-

mencement of a work which promises the greatest benefits to the whole country, not without the hope and prospect of its being the source of still further advantages and improvements. The day being fine added to the pleasures of the scene. Money and cider were distributed to the populace, while the liberal hospitality of Holcombe Court, to the genteeler sort, closed the scene in a manner suitable to the occasion, and worthy the owners of that respectable mansion.

It will be remembered that the Grand Western was linked through its shareholders with the Kennet & Avon, and was soon to be also with the Bristol & Taunton, whose Act was passed in 1811. The energetic John Thomas of Bristol, later superintendent of the Kennet & Avon, also superintended the building of the first part of the Grand Western and was a shareholder in the Bristol & Taunton. It was therefore in the Grand Western's interest to gain influence over the Tone, if their canal were to join the river as the company's Act empowered them, and over the Bristol & Taunton Company. Probably for this reason the Grand Western in the name of John Thomas began in 1811 to buy Tone debt. In 1812–13 they held £4,100 out of a total of £11,150, and in 1816–17 £6,679 out of the £10,733 remaining; then the holdings were transferred to the Bristol & Taunton.

The Bristol & Taunton Act had not been passed, however, when the Grand Western proprietors decided to begin their cutting not at the Taunton or even the Topsham end, but from Lowdwells to Burlescombe on the summit of the main line, and then along the branch to Tiverton. Previously, apparently on the recommendation of John Rennie who had been appointed engineer, it had been decided to lower the original planned level at Holcombe by 16 ft and to lay out the branch to Tiverton afresh and on the summit level, without locks. This was much to be desired, as a large trade in limestone and lime from Canonsleigh quarries to Tiverton was anticipated. Rennie's estimate for completing the Lowdwells–Tiverton section was £86,050. After a year of work, however, the cutting was proving so expensive that a suggestion was made to stop construction, £65,000 having been spent. A General Assembly, after hearing a report that if an Act to raise tolls could be obtained, a revenue of £10,000 pa could be got from the Holcombe–Tiverton trade, decided to go on. That there had been strains behind the scenes is shown by an extract from the report of the meeting dated 15 August 1811:

... this Meeting receives with much Pleasure the Report of the Committee, and perceiving that, although with respect to the whole Canal, only a small Part is effected, yet this Meeting is convinced, that regarding the Time employed, the difficulty and importance of the Work in that Portion of the Canal which has been first put in Execution, the Excellence of the Work thus executed, the Economy in the Consumption of Land, and the quantum of Work done, a laudable Energy is evinced in the Committee of the last Year, and that the Gentlemen who composed it are entitled to the renewed Thanks of the Proprietors.

Soon afterwards the idea of completing the whole canal seems to have been dropped, and instead in January 1812 the committee reports:

The Committee are fully satisfied that if the Canal be executed only between Holcomb and Tiverton the revenue thence arising will be adequate to afford a good interest to the subscribers and that whenever the Canal shall be completed to Taunton a consequent increase of tonnage fully proportionate to the expenses may be confidently anticipated.

Thenceforward engineering difficulties and re-estimates came one after the other as Rennie discovered more and more unforeseen difficulties and necessities. At last, on 25 August 1814, the canal from Holcombe Rogus to Tiverton was open, and 'the first barge arrived by the Canal laden with Coal, which reduced the price 3d. per bushel; thereby producing a benefit to the rich and a blessing to the poor,'[6] and early in 1815 Messrs Dunsford & Browne, and Mr Talbot were advertising lime for sale from the canal wharf at Tiverton.

It may be interesting here to step aside for a moment for a glimpse behind the scenes of canal construction. The *Annual Register* for 1811, under the date 27 April, carried the following item quoted from the *Taunton Courier*:

On Monday last a disturbance of a serious nature occurred at Sampford Peverell. The annual fair for the sale of cattle, &c. was held there on that day. On the Saturday preceding, a number of the workmen, employed on excavating the bed of the Grand Western Canal, assembled at Wellington, for the purpose of obtaining change for the payment of their wages, which there has been lately considerable difficulty in procuring. Many of them indulged in inordinate drinking and committed various excesses at Tiverton and other places, to which they had gone for the purpose above-stated. On Monday the fair at Sampford seemed to afford a welcome opportunity for the gratification of their tumultuary disposition. Much rioting took place in the course of the day, and towards evening a body of these men,

Grand Western Canal.

At a General Assembly of the Company of Proprietors of the
GRAND WESTERN CANAL, held at the City of London Tavern, Bishopsgate-Street, London, on Thursday
the 15th Day of August, 1811, pursuant to the last Ad-
journment.

SAMUEL WOODS, Esq. in the Chair.

The Minutes of the last Meeting, held on the 27th Day of June, 1811,
were read.

A Report from the Committee of Management to this Meeting on the
State of the Works and the Affairs of the Company was presented and read.

RESOLVED, That the same be entered on the Proceedings of this Day.

RESOLVED, That from the Report of the Committee, it appears that the General
Assembly, held on the 27th Day of June, last had not before it the Explanations and Information now afforded, and was therefore juftified in directing special Enquiry to be
made: And that this Meeting receives with much Pleafure the Report of the Committee, and perceiving that, although with respect to the whole Canal, only a small
Part is effected, yet this Meeting is convinced, that regarding the Time employed, the
difficulty and importance of the Work in that Portion of the Canal which has been firft
put in Execution, the Excellence of the Work thus executed, the Economy in the
Confumption of Land, and the quantum of Work done, a laudable Energy is evinced
in the Committee of the laft Year, and that the Gentlemen who compofed it are entitled to the renewed Thanks of the Proprietors.

RESOLVED, That the Thanks of this Meeting be given to the prefent Committee of
Management, for the Diligence and Attention they have fhewn in the Execution of
their important Truft.

RESOLVED, That if the Works be fufpended, a Sum of 75,000l. amounting at leaft
to 25l per Share muft be expended, and from which no Income can be derived.

RESOLVED, That if 75,000l in addition to the above 75,000l be raifed in order to
complete the Line to Tiverton, the Subfcribers will, according to the Report of the
Committee, divide from the contemplated Increafe of Tonnage from 10 to 12 per cent
on the additional 75,000l. or from 5 to 6 per Cent on the whole Capital expended.

RESOLVED therefore, That it appears to this Meeting it will be highly beneficial to
the Proprietors to complete the Line to Tiverton, and that so much of the Refolutions
of the laft Meeting as relate to the Sufpenfion of the Undertaking be rescinded

8. The Grand Western shareholders decide to continue

consisting of not less than 300, had assembled in the village. Mr. Chave (whose name we had occasion to mention in unravelling the imposture respecting the Sampford ghost) was met on the road, and recognized by some of the party. Opprobrious language was applied to him, but whether on that subject, or not, we have not been informed. The rioters followed him to the house, the windows of which they broke; and apprehensive of further violence, Mr. Chave considered it necessary to his defence to discharge a loaded pistol at the assailants. This unfortunately took effect, and one man fell dead on the spot. A pistol was also fired by a person within the house, which so severely wounded another man, that his life is despaired of. A carter, employed by Mr. Chave, was most dreadfully beaten by the mob. Additional numbers were accumulating when our accounts were sent off, and we understand their determination was to pull down the house.

The cost of these 11 miles without a lock had been £224,505, or more than the original estimate for the canal. A one-way trade in coal and limestone grew up, but that in coal diminished after the peace when the fall in the cost of land carriage made it as cheap to bring coal by wagon all the way from Taunton as to transfer it to barges at Holcombe. Instead of £10,000 pa, as had been forecast in 1811, the actual tolls received were as follows:

	£	s	d		£	s	d
1816	574	6	3	1827	739	0	7
1817	unknown			1828	623	15	2
1818	556	13	7	1829	705	2	8
1819	627	5	2	1830	unknown		
1820	752	10	7	1831	756	19	11
1821	632	1	1	1832	918	1	8
1822	unknown			1833	839	18	10
1823	401	16	7	1834	968	9	2
1824	381	1	5	1835	970	11	0
1825	575	3	1	1836	1,164	0	0
1826	700	19	8				

No dividend was paid. In 1818 the committee considered a plan to extend the canal to Taunton on a smaller scale with thirty locks at a cost of £70,000. It was then recorded that 32,000 tons a year of coal and culm were being carried by the Tone to Taunton, and the committee were anxious to make their waterway accessible to this trade. But even if the estimate were reliable, cash resources in land and uncalled shares were only £67,000, and the project dropped.

The revival of the Bristol & Taunton and its conversion to the Bridgwater & Taunton left the Grand Western unmoved. But the latter's opening in 1827 did improve prospects. In August of that year Charles Dean, an Exeter surveyor and engineer, made the first suggestion, that the Grand Western should form part of a small ship canal to replace the now moribund English & Bristol Channels concern. He proposed that a cut should be made from the mouth of the Parrett to Bridgwater, and a dock there; the Bridgwater & Taunton widened and deepened; a new canal built from Taunton to Lowdwells and the old one extended by Cullompton, Bradninch and Stoke Canon to the Exeter Canal at the new basin then being built, with a branch to Crediton. All this should be either for 80-ton craft, including coastal vessels, to cost £500,000, or 100-ton to cost £700,000. It raised no enthusiasm.

James Green was the engineer employed to enlarge the Exeter Canal and build its basin, and it may have been because he had studied Dean's proposals that he induced some of the Grand Western shareholders who lived near Exeter to call a meeting on 1 May 1829, before which he laid a plan for extending the canal to Taunton. He pointed out that the cost of carrying coal to Tiverton from Taunton was too high to make it remunerative for the lime merchant to barge limestone along the canal from Canonsleigh to be burned at the Tiverton kilns, and that the only chance of more traffic was to continue the canal to Taunton. He said that a railway had been suggested, but that apart from the inconvenience of shifting cargoes twice, it would have to be very roundabout, with expensive cuttings and embankments, or it would need inclines with stationary steam-engines. If a canal were chosen, it could be built with locks to the smaller size of the Bridgwater & Taunton for £100,000; but he recommended a tub-boat canal 23 ft wide and 3 ft deep, with two or at most three inclined planes worked by water-power as on the Bude and Torrington Canals. The cost would be about £50,000. Finally, he suggested hopefully that were the Taunton exercise to be a success, the canal could be carried on the same principles to Exeter to join the Exeter Canal. The General Assembly approved the idea 'providing the same can be made and in every respect completed including the purchase of land and damages attending the execution of the work at any expense not exceeding £50,000,'[7] and Green was instructed to make a survey on the basis of the Parliamentary line. At meetings with the Bridgwater & Taunton Company, it was decided that instead of entering the Tone above North Town bridge on the Parliamentary line, the two canals would join each other below the bridge. By the beginning of August he and a local surveyor, John Easton, had completed it.

Green reported in March 1830 that the canal would cost £61,324. There would be seven canal lifts and one inclined plane:

1. Near the commencement of the Canal at Taunton, in order to gain sufficient height to pass the Canal over the Taunton Mill Leat without interference therewith, and also to pass the Canal by an Iron Aqueduct over the river Tone, near Bishop's Hull   20 feet* rise
2. Lift near Norton, or Allerford Brook   16   ditto
3. Ditto near Hillfarrence Brook   19   ditto
4. Ditton near Trefusis Farm, which is of a sufficient height to pass the Canal a Second Time over the River Tone, by an Iron Aqueduct near Ninehead   38   ditto
5. Lift on the North Side of the Tone near Ninehead Court Lodge   24   ditto
6. Lift near Winsbeer Quarry   18   ditto
7. An Inclined Plane near Wellsford   81   ditto
8. A Perpendicular Lift near Greenham, which attains the Summit Level of the Finished Part of the Canal   46   ditto

Total Rise 262 feet[8]

His purpose in suggesting lifts was to save money and water, for locks would have used more of both. He recommended 8-ton boats, four to eight being drawn by one horse, the size of each boat being 26 ft by 6½ ft, drawing 2 ft 3 in. of water when laden. Green himself attributes the invention of the canal lift to Dr James Anderson of Edinburgh. Five experimental types had been built in Britain between 1796, when Anderson wrote, and this time.[9] None had succeeded in ordinary use, and engineers had either gone back to locks or used inclined planes. Green himself probably based his design on the experimental lift at Mells on the Dorset & Somerset Canal described earlier in this book.

A description of Green's lifts by himself will be found in the *Transactions of the Institution of Civil Engineers*.[10] They were not all quite similar, and Greenham lift, the last to be perfected, incorporated the experience gained in making the others work. Each lift consisted of two chambers separated by a pier of masonry. Within each chamber moved a caisson, wooden but strengthened with iron, and having an iron bar running lengthwise above it to which the lifting chains were attached. Each caisson was big

* The heights given were later varied; these are given in Appendix II.

ABSTRACT of Mr. Green's Report of the 2nd of March, 1830.

THE line which I have chosen for the purpose of effecting a Canal communication between the river Tone, at Taunton, and the Summit Level of the Grand Western Canal, near Holcombe-Rogus, will be found by the plans to be a remarkably direct one. It commences at the River Tone, about a quarter of a mile above the Bridge at Taunton, and follows nearly the Parliamentary Line throughout.

This Line is more advantageous to the Landed Proprietors than any hitherto laid down, in several instances it forms a convenient division of property, and by which many bridges (at all times impediments to a Canal) are avoided, as well as heavy expenses in severance damages. It is also more out of the reach of floods, and will be less affected by them than if the original Line were prosecuted.

I have found it necessary, in consequence of the very gradual rise of the land in the first five miles of the Canal from Taunton upward, to make in that distance four perpendicular lifts. I have also made some modification of the respective heights of the several lifts, which have been so regulated as to suit the features of the Ground and diminish expense, whilst the whole number of Lifts will remain the same, viz. Eight, over which a Boat with eight Tons of cargo may be conveyed in twenty-four minutes; and supposing a Horse will generally draw on the Canal four Boats with thirty-two Tons, the time required for the whole ascent of thirty-two Tons of cargo, will be only about one hour and a half. The position and rise of the Lifts laid down in the Longitudinal Section, and as marked on the plan, will be as follows:

1. Near the commencement of the Canal at Taunton, in order to gain sufficient height to pass the Canal over the Taunton Mill Leat without interference therewith, and also to pass the Canal by an Iron Aqueduct over the river Tone, near Bishop's Hull	20 feet rise.
2. Lift near Norton, or Allerford Brook	16 ditto.
3. Ditto near Hillfarrence Brook	19 ditto.
4. Ditto near Trefusis Farm, which is of a sufficient height to pass the Canal a Second Time over the River Tone, by an Iron Aqueduct near Ninehead	38 ditto.
5. Lift on the North Side of the Tone near Ninehead Court Lodge	24 ditto.
6. Lift near Winsbeer Quarry	18 ditto.
7. An Inclined Plane near Wellsford	81 ditto.
8. A Perpendicular Lift near Greenham, which attains the Summit Level of the Finished Part of the Canal.	46 ditto.
Total Rise	262 feet.

The Country through which this Canal will pass is favourable to the execution of such a work, the Ground being generally of easy cutting, and although much Lining and Puddling will be required, the Line of the Canal will furnish abundance of Earth fit for that purpose; good Stone for building may be procured within a moderate distance of the Canal, and in many convenient places good Earth will be procured in, or adjoining the Line for making Bricks. I have, however, designed the Aqueducts Public Road Bridges and Viaducts for carrying Roads under the Canal, (which may be done in many instances, and prove a great convenience to the Public as well as to the Canal) to be principally of Cast Iron; the cheapness of that Article, its convenience in fixing, and its durability, rendering it for these purposes most desirable.

I have also ascertained that the works may be commenced at Taunton and prosecuted upward, whereby the Canal may be made productive in many parts before the whole is finished, particularly at Wellington and some intermediate places.

I have estimated the cost of executing the Work in eight separate sections, which may be contracted for in so many distinct Lots, or by two or more taken together, the total amount of which is £49,324: 6s. 7d.

This estimate has been made with every possible attention and detail, and I have no doubt it will be found adequate to the purpose. The work has been valued at liberal prices, and as the cost of labour and every sort of material is now very moderate, I have good reason to think it may be executed for less than that sum; but as some trifling expenses may occur, which it has been impossible to foresee, it may be prudent to add to the above sum, £2,000. The cost of land and damages cannot exceed £10,000, making a total of £61,324: 6s. 7d.

To the Chairman and Committee of Management   JAMES GREEN.
     of the Grand Western Canal Company.

Wyld, 5, Charing Cross.

9. James Green's report on the proposed extension of the Grand Western

enough to hold one 8-ton boat. At each end of the caisson were vertically-moving iron gates, and there were corresponding gates, one for each caisson, at the ends of the upper and lower pounds. The caissons, when at the bottom of the chamber, rested on wooden cross-beams. Above the masonry of the lift was a cast-iron frame on twelve supports, which held the 22-ft shaft upon which were mounted three iron wheels 15 ft in diameter. These wheels carried the chains that supported the caissons. Gearing connected to the centre wheel enabled the caissons to be moved by hand if necessary.

Normally, however, the lift was worked by adding 2 in, or 1 ton, of water to the uppermost caisson, when it overbalanced the lower. A brake, consisting of a lever working on a brake wheel, then controlled the descent. Because the weight of the lifting chains would alter the equilibrium as the weight of those holding the descending caisson became greater and those holding the rising caisson less as the caisson moved, corresponding lengths of chain were fastened below the caissons to compensate for the changes. When the caisson reached the top of a lift, a hand winch moved a forcing bar which pressed the caisson closely against the gate, shutting off the canal pound. Both gates were then lifted together, and the boat floated out. At the bottom, the descending caisson was forced by iron wedges against the bottom gates.

At the first lifts to be built, Taunton, Norton and Allerford, Green found that the caissons would not sink far enough in the water; he therefore put in gates to form chambers beneath the caissons, and two sets of pipes by which the water in them could be drawn off. These had insufficient capacity, and therefore two locks, each with a 3-ft lift, and each big enough to take a tub-boat, were built side by side just below the caisson chambers. At these lifts also, the iron caisson doors fell flat on to the bottom of the caissons when they were opened. whilst the iron gates of the chambers moved horizontally on pivots. The lower lock gates were of timber, and also moved horizontally.

At Trefusis, Nynehead and Winsbeer lifts locks were built from the start, no pipes having been installed. Here the same machinery raised caisson and chamber doors simultaneously. These also moved vertically, which allowed the locks to be shorter and the water consumption less. Greenham lift had no locks below it; instead, drains were provided adequate to carry off the water from the chambers. Therefore it only took boats 3 minutes to pass Greenham lift, against 8½ minutes at the others.

After their teething troubles, mostly caused by faulty ironwork or careless operation, Green's lifts appear to have worked well as long as the canal lasted. The Anderton lift between the River Weaver and the Trent & Mersey Canal, opened in 1875, is the only other fully-operational lift to be built in this country; abroad, large-scale examples are at work in several countries.

A special assembly in London approved Green's report and proposals at an expenditure not to exceed £65,000. The committee thought the net profit would produce 10 per cent on the outlay, though some proprietors hoped for fifteen. On Monday, 13 June 1831, work began near the proposed junction with the Bridgwater & Taunton, where there was to be a 'handsome, lofty aqueduct over Rowbarton Road', and between Holcombe Rogus and Wellington.[11]

The committee report of 28 June 1832 was not cheerful. They had tried to sell shares forfeited during 1810–14, the buyer to pay £21 for the calls now due,* but had had no bids. Arrangements had therefore been made to distribute them free among the existing shareholders on condition that the calls were met. They apologized for the delay that was taking place in construction, which they attributed to opposition from a body of turnpike trustees. There was to be more. The committee, reporting to the General Assembly of 25 June 1834, said no one thing was the cause, but that

it must rather be attributed to an accumulation of petty difficulties which a supineness and want of activity and foresight on the part of the Contractors have not enabled them to surmount, notwithstanding they have been urged forward by repeated remonstrances, both from the Committee, and the Engineer. With some difficulty they have been roused at length to exertion, and have signed an agreement, under heavy penalties, to complete the first three Lots, as far as Bradford, by the 1st. August, and the whole Line by 1st. October next.'

The engineer now reported that it would be necessary to have locks at Taunton and Lowdwells to prevent any variation in the water levels of the Bridgwater & Taunton or of the summit altering those at the endmost lifts, the working of which depended upon an unvarying level. As built, that at Taunton was in fact a stop-lock. That at Lowdwells normally had a rise of 3½ ft, and could take four tub-boats at once.

* The £100 shares were £79 paid up, and the committee hoped to complete the canal by calling the remaining £21. The same took place on 2 March, and 360 shares were offered.

Grand Western Canal;

360 FORFEITED SHARES,

(By Order of the Committee.)

SHARES IN NAVIGATIONS,

Covent Garden Theatre, Gas Companies, &c.

The Particulars

OF

SHARES

IN THE

Grand Western Canal;

IN THE

Grand TRUNK, LANCASTER, STRATFORD, THAMES, and SEVERN, and other NAVIGATIONS;

IN

COVENT GARDEN THEATRE,

CITY of LONDON and WESTMINSTER GAS COMPANIES,

AND

London University;

WHICH WILL BE SOLD BY AUCTION,

By Mr. SCOTT,

(Nephew of the late Mr. T. SCOTT,)

At the MART, opposite the BANK,

On FRIDAY, the 2d Day of MARCH, 1832,

At Twelve O'Clock, in Lots,

(IF NOT PREVIOUSLY DISPOSED OF.)

Printed Particulars may be had at the *Mart*, and of *Mr. SCOTT, Estate and Canal Agent,* 8, CAREY STREET, LINCOLNS INN.

N. B. On the same day will be sold, a *Freehold House*, and *Premises*, producing £40. a Year, well secured, situate at *Clerkenwell*, near the intended *New Street* from *Holborn Bridge.*

---

10. Auction advertisement for the sale of forfeited Grand Western shares

In spite of the agreement with the contractors, the canal from Taunton to Bradford was not opened until 24 February 1835, when

> seven barges, laden with coals, attended by another barge in which were a band of musicians and several spectators, yesterday passed through the aqueduct which crosses the Kingston road,* having been brought into that higher level by the novel and efficacious process of lifting.[12]

When the committee once again make their annual report in June, it is not now upon turnpike trustees or an accumulation of petty difficulties that their blame falls for the delay in opening the whole line, but upon James Green the engineer:

> the true cause appears to rest in the novelty of the plan of the Lifts, and the want of foresight as to actual inconvenience which might arise in using them. Although in January, 1834, the Lift near Taunton was reported to be ready and to answer every purpose, Mr. Green relying with too much confidence on theoretical principles, never subjected it to a full and fair trial, so that many practical difficulties were only gradually developed and detected, which had not in the first instance been either observed or anticipated.

The canal was, however, open with five lifts to Wellington, and since Green had agreed to build the lifts for a fixed sum, he had to bear the cost of alterations, though the company suffered from delay.

The next year, 1836, came, and still the canal was not open. Instead, the lift at Greenham at the junction with the old section, with its rise of 42 ft, the highest on the canal, had partially collapsed owing to the subsidence of the earth banks, while the inclined plane at Wellisford failed to work at all. On 27 January the canal company announced that Green had ceased to be engineer, his place being taken, with the title of superintendent, by Captain John Twisden, late of the Royal Navy. The company also called in W. A. Provis, an experienced canal engineer and contractor, to report on the lifts and plane. He found the former satisfactory, the latter not.

It is most curious that Green, who had planned and built the Bude Canal with six and the Torrington Canal with one inclined plane, and who must have had many chances to see them at work and experiment with them, should have failed to build the Wellisford incline correctly. His failure to do so probably explains why he was not chosen as the engineer of the Chard Canal, the Act for

* At Taunton.

which had been passed in 1834 following a survey and estimate by Green.

Wellisford was a double track plane, with a slope of about 1 in 5½. It was 440 ft long, with a rise of 81 ft. Two cradles ran on lines of rails from the upper to the lower pounds of the canal. If a loaded boat were passing in each direction at the same time, then the motive power needed to overcome the friction of the machinery was small; but if a loaded boat were to be pulled up when the down-coming cradle was empty, the motive power needed was much greater. This motive power was provided by water. Two deep wells were dug, and in each was hung a bucket that could hold rather more than 10 tons of water, one being at the bottom when the other was at the top. To work the machinery, water was led into the uppermost bucket till it counterbalanced the weight of the ascending boat and cradle on the incline and the weight of the empty bucket. When the full bucket reached the bottom, a valve was forced open which released the water into a drain.

Green, though he had built the Hobbacott Down plane of the Bude Canal on the same principle, failed to provide enough power. Provis went to look at Hobbacott Down, where 15-ton buckets were used to raise 4-ton boats, and considered that to raise the 8-ton Grand Western boats 25-ton, rather than 10-ton, buckets ought to have been provided.

The situation of the canal company was now desperate. Receipts were low, because the line was not yet open; money was exhausted, because the share capital had been fully subscribed, and mortgages had been incurred as well. They borrowed over £1,000 from Captain Twisden, tried and failed to borrow from the Exchequer Bill Loan Commissioners, and passed the hat round the members of the committee. There was a move early in 1838 by some gentlemen connected with the Bridgwater & Taunton Canal, who offered a loan, but this fell through. There was talk by Twisden of building thirty locks at Wellisford on a novel principle instead of the plane; the leader of the Bridgwater & Taunton party described him as one who 'knows nothing of engineering and is 80 years of age'. At last the hat was again passed round the committee and the superintendent, and enough was raised to repair the Greenham lift and buy for £800 a steam-engine to work the plane. On 28 June 1838 the canal was opened, and the local newspaper, whose reporter had not fully digested his science, wrote as follows:

We have unfeigned pleasure . . . in announcing that the trade on this canal which has been suspended for nearly four years, from the bad arrangement of the engines, will be immediately resumed in consequence of the adoption of a scientific process by which this and every other canal in the kingdom may be successfully put in operation at a trivial expence, compared with the former elaborate, costly, and inadequate modes of working over the levels. This has been effected by a simple steam engine, of twelve horse power, which entirely saves the waste of water so exhausting to the practicability of most canals. The chief obstacle to the Grand Western has been the *Welsford plane* . . . It is peculiarly gratifying to be able to state that the disheartening difficulty which has so long prevailed . . . has been obviated by the scientific skill and industry of our townsman, Mr. James Easton, son of the worthy veteran engineer and land surveyor, Mr. Josiah Easton, Sen. of Bradford. Mr. Easton's engine was set to work last Thursday, on a boat and load weighing seven tons . . . and effected its object with ease, to the astonishment and admiration of all present, in *five minutes*. . . .[13]

The objective of connecting the Channels had long ago given way to the limited aim of supplying the country between Taunton and Tiverton. Coal could now be brought from Bridgwater, and it was probably at this time that kilns were built at Holcombe Rogus, so that lime could be supplied to the country between there and Taunton. Cargoes had, however, to be transhipped at Taunton, because tub-boats could not work on the Parrett, until the opening of the Bridgwater & Taunton's extension to Bridgwater dock in 1841. Boats usually worked in gangs of four, six or eight drawn by one horse.

Left to itself, the canal might have proved a modest success for its long-suffering shareholders. The tolls received in the first few years of the extension were as follows:

	£	s	d		£	s	d
1836	1,163	19	11	1841	3,631	19	7
1837	1,822	19	9	1842	4,114	15	1
1838	2,754	4	0	1843	4,768	15	11
1839	not known			1844	4,925	10	8
1840	3,456	19	3				

After 1840, however, the figures were helped by the tolls charged on the carriage of railway building material, for the last chapter of the canal's history was to begin. On 1 May 1844 the Bristol & Exeter Railway opened for traffic, and for 1845 the Grand Western's receipts had fallen to only £2,819 6s 6d.

Even the prospect of this fall had frightened shareholders, for at the end of 1845 the railway company's minutes record:

Mr. Smith the Secretary of the Grand Western Canal Company had an interview with This Board and presented some particulars respecting that property. He was informed that the subject should have due consideration and that the Board would be willing to receive a proposition as to terms. He left a plan and sketch of the Canal. It was agreed to confer with Mr. Brunel upon the desirableness of entertaining a proposal of this nature.[14]

Perhaps Mr Brunel did not think it desirable, for no immediate action followed. These figures summarize the history of the canal for the next few years.

Date	Tolls Received			Total Tonnage	Through haul Taunton–Tiverton	Receipts
	£	s	d	Tons	Tons	per Tons
1846	2,819	6	6	50,326	12,089	0·93
1847	2,291	15	11	56,512	10,532	0·81
1848	1,735	8	6	37,417	2,456	0·93
1849	2,351	18	4	29,776	2,107	1·58
1850	1,660	8	3	30,047	1,961	1·1
1851	890	5	2	33,552	2,282	0·53
1852	974	19	11	37,354	4,374	0·52
1853	956	1	11	35,754	1,579	0·43

In September 1847 the Bristol & Exeter shareholders were told that the works on their Tiverton branch had been delayed by the difficulty of making any reasonable arrangement with the Grand Western company. But agreement must then have come quickly, for soon afterwards the canal was closed for some weeks whilst the aqueduct was built that took it 40 ft above the track, £1,200 being paid to the company in compensation. This was a cast-iron duct supported on two cast-iron arches, the whole being enclosed in brickwork and the intervening spaces then puddled.[15]

On 12 June 1848 the Tiverton branch opened. Back in March the railway's deputy chairman seems to have proposed that, to prevent rate-cutting, the canal alone should carry lime, and that coal should be brought from Bridgwater by rail to an interchange siding at Taunton to be sent thence by canal to Tiverton, the siding and canal bank to be used free by both companies. This proposal was agreed to in May, the siding being built in June. The Grand Western were to charge 1½d a ton per mile on rail-borne coal on their canal, but the full parliamentary 3d on that off the Bridgwater & Taunton. But when the branch opened, the price of coal at Tiverton wharf rose sharply, and there were complaints

IX. Aqueducts on the Grand Western Canal: (*above*) over the Tiverton branch near Halberton; (*below*) over a private road at Nynehead near Wellington

X. (*above*) Clay cellars at the head of the Hackney Canal in 1955; (*below*) the Stover Canal at Teignbridge in 1954

of monopoly loud enough to bring it back to its old level by mid-July.[16] This toll agreement evidently accounts for the improvement in the canal receipts per ton carried in 1848, 1849 and 1850.*

A change dates from a letter received by the railway from Henry and Charles Fox of the Wellington Mills advocating lower rates, and saying that the competition of the canal was not to be feared. This letter reached the railway on 29 December 1849, two days after Mr Smith from the canal company had met the directors to support the existing system of agreed charges. Five months later the first hostile proposals were made to the railway board, to carry goods from Bridgwater to Wellington at the same charge as to the canal siding at Taunton, and soon afterwards to close the siding. Mr Harriott, the railway official who made these suggestions, was evidently in favour of ending canal competition, for in the same month of June 1850, he seems to have tried to bring about an agreement between the railway and the Bridgwater & Taunton Company.

He did not succeed, but early in 1851 the railway made a new and less favourable rate agreement with the Grand Western. The siding was now out of use, and the railway was carrying its own coal, but in return for the canal company charging 3d per ton per mile on coal and salt entering their canal from the Bridgwater & Taunton, the railway agreed to maintain fairly high rates for coal, salt and lime from Dunball near Bridgwater: 3s 6d a ton to Wellington, 5s to Westleigh (Burlescombe), and 4s 9d to Tiverton against 4s 6d to Cullompton and 3s 9d to Exeter in their own trucks.

Judging by the figures given earlier, the result was a heavy drop in the canal company's receipts. In March 1852 they gave notice to the railway to end their arrangement with them, and a few weeks later agreed to join the Bridgwater & Taunton in cutting tolls and fighting the railway, who at once decided to retaliate. Whereas under the agreement of 1851 coal had been carried from Dunball to Tiverton by rail in company's trucks at 4s 9d, it was on 26 August 1852 *reduced* from 1s 9d to 1s 3d a ton. A few days later it was reduced again, together with lime and brickyard goods, to 1s a ton, and lime from Westleigh (Burlescombe, also on the canal) to 3d a ton to Tiverton or Wellington. In October coal was being carried from Bristol to Tiverton at 3s 6d a ton. Meanwhile the

* *White's History, Gazetteer and Directory of Devon,* published in 1850, says: (the canal) is worked on friendly terms with the railway, under the able management of H. J. Smith, Esq.

H

canal company had been carrying coal at ¼d per ton per mile and for a time free of toll, while the inhabitants rejoiced in the benefits of competition. The results in tonnage and receipts can be seen in the figures already quoted.

The price cutting had its effect, but first, as we saw in Chapter V, the canal pass was to be sold by the Bridgwater & Taunton when they made a proposal in August 1853 to the railway that in effect suggested that the Grand Western should be bankrupted in exchange for a monopoly to the Bridgwater & Taunton. The proposal caused consternation among the Grand Western's directors, for without the Bridgwater & Taunton's co-operation in charging low tolls they could not obtain the traffic, mainly coal, with which to compete with the railway. They saw that the end had come, and at the same railway board meeting in October that approved the Bridgwater & Taunton's proposals, H. J. Smith, the Grand Western's superintendent, attended with that company's proposals. These offered a lease in perpetuity at £2,000 pa, and an option to purchase within 21 years. Of the £2,000, £600 pa was to be paid to Smith to run and repair the canal. These terms were agreed with minor changes.

The balance of £1,400 pa was distributed to the shareholders as a dividend, the first they had ever received. The railway company, on the other hand, farmed the canal tolls to Mr Smith for £500 pa. He seems to have come out on the right side on his maintenance agreement, but evidently failed to pay his way on the second, for in 1855 his payment was reduced to £250 for 1856 and £200 pa thereafter.

In 1856 the railway company recorded the effect that the price-cutting had had upon them: the following figures can be set against those of the Grand Western already given.

There can be no doubt but that the arrangement of the Bristol and Exeter Railway Company with the Grand Western Canal Company has been very advantageous to the former; we find that by Mr. Harwood's accounts that coal, culm and lime carried by the Bristol and Exeter Railway from Taunton to Hele has increased in Tonnage and greatly in Toll and Dues, shown as follows, viz.

"Carried from 11th. Oct. 1852 to 11th. Oct. 1853	46,756 tons	£3,749 10s 6d
Carried from 11th. Oct. 1853 to 11th. Oct. 1854	50,992 tons	£10,324 16s 5d
Carried from 11th. Oct. 1854 to 11th. Oct. 1855	53,821 tons	£10,863 3s 3d."

These figures are important because they prove a very large increase in the Revenue with comparatively but a small expenditure in the cost of Locomotive Power and wear and tear of the Line.[17]

In 1864 the option was exercised, and the concern sold for £30,000, H. J. Smith being paid £600 compensation. The purchase was authorized by an Act[18] of the same year. The canal was transferred on 13 April 1865, and in 1867 the section between Lowdwells lock and the Bridgwater & Taunton was closed, the plane and lifts being dismantled. The ¼-mile section west of Lowdwells lock was allowed to go dry, the remainder to Tiverton being retained for stone traffic.

Receipts for the years 1869–73 averaged £382; between 1877 and 1892 they varied between £137 and £303. In 1890 tonnage was 4,539, all of it limestone. The canal suffered from chronic leakage through fissured rock near Greenway bridge, Halberton, and to the end of its life this section has had to be closed from time to time for repuddling. The stone traffic ended about 1924, and in 1962 the remaining portion of the canal was abandoned by the British Transport Commission.

CHAPTER IX

Other Canals of Devon

Exeter & Crediton Navigation

THE Exeter & Crediton Navigation was first put forward in 1792 by a group of Exeter men headed by Richard Chamberlain and John Pinhay, who proposed a line from Four Mills (approximately where Crediton station now is) to the Exe at the public quay just below Exe bridge, which was surveyed by Thomas Gray, surveyor to Exeter corporation. The group seem to have kept the project very much to themselves, so much so that on 5 February 1793 a meeting was held at Crediton to protest against the exclusion of the public, and against such a limited scheme. It proposed instead what became the Public Devonshire Canal project.

That failed about 1795, and probably preoccupation with the war prevented any action being taken on the earlier scheme. But in 1800 Robert Cartwright surveyed it preparatory to going to Parliament, and in 1801 the line originally proposed, from Four Mills to the public quay, was authorized, Richard Chamberlain and John Pinhay being named in the Act. The company's capital was to be £21,400, with power to raise £10,700 more; Exeter corporation were subscribers. Land improvement was a main motive, and tolls not exceeding 2d per ton per mile on the Exeter Canal for lime and limestone were stipulated for cargoes also passing on the Exeter & Crediton, unless the latter company charged more themselves.

There was to be an entrance lock from the Exe into a basin, whence the line was to run round the weirs above Exe bridge, then past Exwick partly in the river and partly by canal to Cowley bridge, and up the Creedy valley, alongside or in the river. The length would have been about 8 miles.

Except for the issue of some share certificates after a general meeting early in 1802, the scheme lay dormant until mid-1808, when a start was made. Land was acquired, including some from Exeter corporation for a towpath from the Exeter Canal to the public quay, mills at Newton St Cyres were bought, and in 1810 about ½ mile of canal was cut roughly from above Exe bridge under the Western Region main line towards Exwick; a portion of this is shown as 'Old Canal Road' on the deposited plan for the South Devon Railway about 500 yards west of Exe bridge, and in enlarged form still exists as a flood relief channel. Then work stopped, though a call was made in July 1812. In 1818 the scheme was abandoned, and in October 1820 unwanted land was leased back by the company, who seemingly ceased to function after 1822.

The Public Devonshire Canal project

What became the Public Devonshire Canal project arose out of a meeting held at the *Ship*, Crediton, on 5 February 1793 to protest against the proposed Exeter & Crediton Navigation as not being open to public participation, and so limited as to be of 'infinite prejudice to the public at large'. Instead, it was suggested that a canal should start at Coleford near Colebrooke, and run past Crediton to join the proposed Grand Western Canal near Topsham. Some of those present thought that the suggested scheme might be extended still farther inland, to Bow, North Tawton, or elsewhere, and 142 gentlemen promised to subscribe £67,600.

A committee was then appointed. These reported to another meeting at Crediton in October,[1] recommending a canal from Topsham to Barnstaple, so connecting the English and Bristol Channels, with branches to Exeter and North Tawton. Their engineer, George Bentley, and surveyor, Thomas Bolton, had worked out a line which Robert Whitworth had approved. As printed, it showed a main line 48¼ miles long, rising 287 ft from Topsham by Broad Clyst to Stoke Canon and Cowley bridge, then by the Creedy valley to the beginning of the short summit level at Copplestone. This ended at Bradiford, whence the line fell 312 ft down the Taw valley to Barnstaple. The Exe branch came off at Cowley bridge, and fell by 57 ft in 3⅝ miles. It crossed the Exe below Exwick and finished below Exe bridge. That to North Tawton left the main line at Down St Mary and rose 55 ft in 8 miles. This plan showed no physical connection with the Exe or the Grand Western Canal at Topsham, in view of the uncertain prospects of that scheme, or with the river at Barnstaple.

The committee estimated a canal 28 ft wide at surface and 4½ ft deep, to take 25-ton narrow boats, to cost £195,000, on which goods could be carried at 2½d per ton per mile, toll and freight, against 1s on the roads. The traffic, they thought, would be mainly manure from the ports (they meant lime and sea-sand from Barnstaple Harbour), and corn for export. They also saw it carrying Irish wool and yarn imported at Barnstaple to the woollen mills; clay, cider and timber; coal brought in at either end; and, probably because of the culm workings at Tawstock near Barnstaple, hoped also that coal and iron would be found near the canal. Optimistically, they reckoned income would be £11,650, per annum and wrote:

> To these (canals) the increased energy and prosperity of the North of England ... may to a great degree be attributed; while this Western part of the Kingdom (almost destitute of them) has declined in nearly an equal proportion; from the rivalship of the North, the woollen manufacturers of the West have peculiarly suffered.

By 1795 the plan had been dropped, though in November 1800 it came briefly to life to protest against the Exeter & Crediton project, and in 1824 there was talk of it in the form of a line from Weare Giffard above Bideford to Torrington (this was when the Torrington Canal was being built) and on to Lord Clinton's seat at Huish; then across to the Taw valley at Eggesford, and by Exeter to Topsham, heavy goods from London to north Devon, and sea-sand for manure in return, being thought of as traffics.[2] In 1831 it was still being mentioned.[3]

River Teign, Stover Canal and Hackney Canal

The Teign estuary was very different from the Exe. Until the beginning of the clay traffic in the 1740s, there were no places along its banks or at its head that could generate much traffic for its only port, at Teignmouth.[4] When the clay trade began, most of it was taken from the country round Heathfield to the shores of the Teign and shipped in barges carrying about 30 tons each to be loaded on board ship at Teignmouth.

Until about 1760 the trade was small; then it increased rapidly, clearly providing the incentive for the building of the Stover Canal. Here are some figures:

	Tons
1740	385
1758	1,648
1765	2,947
1774	3,626
1784	9,428
1789–92 (av.)	12,072

James Templer, of Stover House, Teigngrace, began the canal in January 1790 as a private venture, seemingly intending to carry it by way of Jewsbridge, Heathfield, to Bovey Tracey, with a branch to Chudleigh. He built it to Ventiford, his engineer being Thomas Gray of Exeter, and then in 1792 obtained a private Act[5] to enable him to raise money by mortgage to continue it. However, he did not use its powers, and clay cellars were built at Ventiford.

The canal is $1\frac{7}{8}$ miles long, with a staircase pair of locks near the entrance, and three others above. The three lower locks were originally earthen-sided, but were later given brick or timber chambers. All (except Graving Dock lock) could take two 54-ft barges at once, end to end; the lowest was enlarged in 1841 to 215 ft × 45 ft, presumably so that it formed a basin where craft could wait for the tide before leaving the canal. The canal joined the Teign at Jetty Marsh, Newton Abbot, whence barges passed through the dredged Whitelake channel to the estuary, and so down to Teignmouth. Newton's own quays were at Jetty Marsh and in the channel.

Soon after it was opened, a local author wrote of the clay that it was used

> for the fine earthern manufacture of Staffordshire: it is dug in the parish of King's-Teignton only, formed into little irregular squares somewhat like bricks, weighing thirty-six pounds each. About seven thousand tons are annually sent from these pits, in barges on Mr. Templer's Canal, to the river Teign, and down that river to the port, at 1s. 6d. per ton, toll and boatage. . . . This little Canal . . . is the means of great benefit to the inland country, by the return of coal, and other necessary and useful commodities. Great part of the barren waste of Bovey Heathfield, and the lands near his elegant seat, appear in consequence, like a *new creation*, from the wide chaos around it.

Other local minerals were exported from time to time; imports were coal, culm, limestone and sea-sand for manure.

In 1795, soon after the canal was opened, seventeen barges were working on the estuary; ten of them, including the biggest, of

35 tons, worked up the canal, seven being Templer's. These were either sailed or bow-hauled by men. Frank C. G. Carr, in his book *Sailing Barges*, says that the rig used was the last surviving example of the old Norse or Viking rig, and gives an illustration.

The trade of the Teign increased rapidly thenceforward, from about 400 barges a year in 1790 to 600 in 1816 and 1,000 in 1854. Clay exports from Teignmouth rose too; here are representative figures:

	Tons
1800	15,252
1815	23,186
1838	19,090

Thirty years after its opening the canal was still being used to export clay. It also had a new traffic. The Haytor granite tramroad was probably built as a result of George Templer, James's son, having received a contract for some of the stone required for the new London bridge. Opened on 16 September 1820 from his quarries at Hay Tor to Ventiford, it was then about 7 miles long, but was later extended to about 9 miles, or 10 including sidings. The gauge was 4 ft 3 in, and the rails were made, not of iron, but of granite slabs about 15 in wide, with inside kerbs 2 or 3 in. high. When the Moretonhampstead & South Devon Railway was built, the tramroad was shortened to join it at Granite Siding, 1 mile south of Bovey.

In 1825 George Templer formed a Devon Haytor Quarries Company, which was given contracts for granite used in building the old General Post Office, the British Museum, and the National Gallery. In 1827 Templer also built the New Quay at Teignmouth, so that granite and clay could be loaded there instead of being transhipped in midstream. In the latter year, as the result of a meeting called at Newton in December 1826, James Green was employed to consider how the approach to Newton could best be improved. He proposed a canal about a mile long, with a tide-lock at its entrance, running across shallows to a basin in the centre of the town. A plan was deposited, but no action followed. Two years later, a Devon newspaper reported a project for a canal from Newton Abbot to Torquay.[6]

In 1829 George Templer, short of money, sold the canal and tramroad to the Duke of Somerset, who in turn leased the canal. The clay and granite trades both being prosperous, in 1836 an Act, promoted among others by the Duke, was passed for improving the harbour of Teignmouth and the navigation of the

Teign up to the junction with the Stover Canal. It set up Commissioners with power to borrow £12,000 and take tolls on the river, one of their tasks being to dredge the river to the depth of the sill of the Stover Canal tide-lock. The busy clay trade also encouraged the building by Lord Clifford of the Hackney Canal, 5 furlongs long with a sea-lock, opened on 17 March 1843. It runs from the River Teign a short distance east of the Stover Canal to the Newton Abbot–Kingsteignton road. The Commissioners did in fact improve the river by dredging and staking a river channel. Upwards from Netherton Point this split into two, one running to the Hackney Canal, the other to Newton quays and the Stover Canal.

In 1844 the South Devon Railway was authorized from Exeter by way of Teignmouth to Newton and Plymouth, and Sir John Rennie was retained by the Teignmouth Harbour Commissioners to make sure that its embankment did not affect their navigation, and that big enough bridges were built at Newton over the river and the Hackney Canal. At this time George Davis was running a boat for passengers and goods between Teignmouth and Newton, but when the railway opened he took it off.[7] In January 1845 the inhabitants of Newton asked the South Devon Railway for a station nearer the town than had been planned, which would also be more convenient for the wharves, and this was agreed.

In 1858 the Moretonhampstead & South Devon Railway was promoted (as the Newton & Moretonhampstead). This needed to pass through the Duke of Somerset's land. In the course of negotiations the railway company bought the canal in 1862 for £8,000, with no obligation to maintain it above Teignbridge, and also a little over a mile of the tramroad needed for the line. Although the Hay Tor Railway had been disused since 1858, the Duke insisted that a siding and crane should be built where the rest of the tramroad diverged from the railway; it was probably never used.

The line was opened on 4 July 1866. A little afterwards, the existing lessees, Whiteway & Mortimer, gave way to Watts, Blake & Co., who sank and owned clay-pits. They at first paid £460 per annum, but this fell to £330 per annum from 1869. The active length was now 1¼ miles to Graving Dock lock. Watts, Blake and their successors continued to lease the canal on varying terms from the railway companies, first the South Devon and then the Great Western, until 1942.

The clay trade had gone on increasing. It reached 50,000 tons in 1868, 60,000 in 1878, 80,000 in 1896 and 100,000 in 1905, most of this tonnage going by water. In the early part of this century, before the first world war, between 2,000 and 3,000 barge-loads a year moved down the Teign, under sail or behind a tug. Then road transport began to take over; the water-borne trade declined, and so barges wore out and bargemen became scarce. The Hackney Canal ceased to be used in 1928. In 1931 sixteen barges were still at work, but by 1939 the Stover Canal also was disused.

Ashburton Canal project

In 1792 a canal was proposed from the navigable River Dart above Totnes bridge up the river valley to Ashburton. F. King surveyed it in 1793, and proposed a line rising 202½ ft by 24 locks to a basin below Ashburton, to cost about £15,000, and a plan was deposited. However, there were second thoughts, and George Bentley was called in. He and King now seem to have got the cost below £10,000, and a meeting in January agreed to seek a bill for the revised proposal, firmly enough to make the trustees of the North End, Totnes, turnpike road call a meeting to consider whether to oppose it, as their tolls seemed likely to be reduced. But the idea, caught up in less favourable times, was then dropped.[8]

Cann Quarry Canal

In 1778 Smeaton[9] was asked by John Parker to survey for a canal from Cann slate quarry to the new bridge over the Plym at Marsh Mills, whence barges could come up on the tide from Plymouth. He considered it practicable to build a 2¼-mile long canal of 12 ft bottom width, falling 30 ft by locks to the tideway, but he thought that to make it economic a greater traffic was necessary than the quarry's likely output. He therefore suggested a railroad instead, at about half the cost. But no action seems to have followed, though the Parkers, ennobled as the Earls of Morley,* continued to work the quarry.

The Plymouth & Dartmoor horse tramroad was authorized by Acts of 1819 to 1821, and was built from Princetown to Plymouth. In order to get the Earl of Morley's consent to the deviation authorized in 1821, three of the directors on behalf of the company agreed with him 'for making, at their or the Company's expence, an inclined Plane or branch Railway, communicating with Cann Quarry . . . and for procuring a responsible Tenant'.[10] The financial difficulties of the company prevented anything being

---

* John Parker, born in 1735, was made Lord Boringdon in 1784. He died in 1788, being succeeded by his son John, created Earl of Morley in 1815.

done, until in 1825 the Earl began proceedings against the three. The company then proposed to him (he was himself a member of the committee) that if he made 'such Branch Railway or Communication', they would charge a low rate of toll on their own line for goods coming from it.

The Earl did not commit himself, but went ahead to build a small navigable mill leat, 6 ft wide, from the quarry in the Plym valley down to Marsh Mills, and then wrote to the company:

Doubts are now pending whether it will be best to convey the produce of the Quarry at Cann to the port by completing the canal, now in great part made and connecting it with the Lary* by a Tunnel under the Turnpike Road, or by uniting (by means of a private Railway) the Canal (as far as it is made) with the Plymouth and Dartmoor rail road at Crabtree.[11]

A few days later the Earl and the company had agreed on tolls, which had been the object of the Earl's letter, and he wrote:

my proposition having been acceded to, a junction betweeen the new Cann Canal and the Plymouth and Dartmoor Railway will forthwith be effected and will take place at Crabtree.[12]

Moore, in his *History of Devonshire*,[13] records that 'A new canal and rail-way, communicating from the Cann slate works to the Catwater, were opened on the 20th November, 1829, by the passage of boats and waggons containing large quantities of paving stones and slate from the quarries.' The date of 20 January 1830 is given in the *British Cyclopaedia* (1835), but as the minute book of the P. & D.R. for 5 January 1830 proposes a new toll for slate rubble 'in order to create a trade from Cann Quarry', I think the earlier date the more likely. The communication at this time therefore consisted of a tub-boat canal just under 2 miles long from Cann Quarry to Marsh Mills, and about half a mile of tramroad crossing the River Plym on a bridge with two cast-iron arches to join the main P. & D.R. line.

At some time before 1839 the canal was superseded by an extension of Lord Morley's tramroad upwards alongside the canal to Cann Quarry, presumably to save transhipment. The waterway then seems to have been used only as a mill leat.

In 1852 the South Devon & Tavistock Railway was being promoted to build a locomotive line up the Plym valley to Yelverton and Tavistock, and its representative agreed with Lord Morley and the Lee Moor Porcelain Co. to build a branch line to Lee Moor. The main line was to run parallel to the tramroad, and the branch was to leave at Plym bridge just below the quarry. This

* River Laira, the tidal estuary of the River Plym.

branch was built first and completed in 1854, the year that the main line was authorized. It was transferred to Lord Morley in 1856. The quarry closed in 1855.

Tamar Manure Navigation

The River Tamar had been navigated as far upwards as Morwellham, the quay for Tavistock, since the twelfth century.[14] The tide ran somewhat higher than this, to a fish-weir Lord Mount Edgcumbe had built half a mile below Newbridge, Gunnislake.

When the canal mania came to the west, a meeting at Tavistock in February 1793 discussed a possible line to run from the Tamar via Tavistock to the proposed Public Devonshire Canal (then seen as running via Crediton to Okehampton), with links also to Launceston and Hatherleigh. A rough survey showed it to be feasible, and a second meeting in March, which was attended by a deputation from the Public Devonshire, welcomed the possibility of carrying goods from London, Bristol, the north of England and Liverpool to Plymouth, and opened a subscription list. Support seems to have come not only from Tavistock, but from Ashburton and Exeter; committee meetings were sometimes held at Ashburton, and for a time the company's offices were at Exeter.

Bentley and Bolton were now engaged to survey and estimate the line. As they were also involved at the time with the Public Devonshire and Bude proposals, they were in a position to co-ordinate them if necessary. But the Public Devonshire soon faded away, and their proposal of 1794 was for a line from above Morwellham quay on one level to Wrixhill (Dunterton), and then on for $4\frac{1}{2}$ miles to Launceston by way of an inclined plane down to an aqueduct over the Tamar. There was to be a $9\frac{1}{8}$-mile feeder to the River Lyd near the gorge, and a level branch from near Collacombe to Tavistock via the Lamerton and Langford valleys, with two tunnels. For the time being a metalled road 924 yards long would join the end of the canal above Morwellham to the river, but this might later be replaced by locks. Their estimate was £74,096, and the traffic envisaged some 16,000 tons a year of lime, coal, culm, timber, bricks, stone and slate landed at Morwellham quay and carried upwards.

Bentley and Bolton were asked to think again, and this time they proposed a different line which avoided much rocky ground. The Tamar was now to be made navigable from Morwellham upwards to Blanchdown a little above Newbridge on the Devon side, whence a canal was to run to Tamerton bridge, rising 272 ft as it did so, with a branch over the Tamar to Ridgegrove mill, Launceston. The Tavistock branch was dropped.[15]

These proposals then went to John Rennie, who employed James Murray to resurvey the line. It was clearly now envisaged that at Tamerton bridge or Launceston the canal would join the Bude, for his preliminary report in August 1795 caused the committee to say there was

> every well founded hope, from the inexhaustible lime rocks at or near Plymouth, and from the valuable sea-sand near Bude, of which there is likewise an inexhaustible store, that such advantage will be derived as to justify the most sanguine expectations both of the adventurers and the public.[16]

Rennie approved the scheme in December 1795. For the river section he proposed a 5-ft depth, that which existed between Morwellham and Nutstakes (where the lock was later to be built), except for a few shoals, so that serious dredging would only be needed upwards to Blanchdown. He estimated the total cost at £80,803, supporting in evidence before the Lords Committee in 1796 'the Practicability and great Public Utility of the Undertaking'. Some £40,000 had been subscribed at the time of the Act.[17] This empowered the company to make the Tamar navigable from Morwellham to Blanchdown, though no quays were to be built on the Devon side below Newbridge. A basin was authorized near Blanchdown House, and a tub-boat canal thence to Tamerton bridge, with a branch to Ridgegrove mill, the locks of which were to be between 5 ft and 9 ft 9 in. wide, and multiples of 12 ft 6 in. in length. The authorized capital was £81,000, and £20,000 more if required. Because the salmon fisheries and fish house belonging to Lord Mount Edgcumbe and the Duchy of Cornwall might be affected by the permanent weir which was to replace the old fishery weir at Weir Head, the company had to pay £200 pa for them. Moore's *History of Devonshire* says that this sum was greater than the rents received.

By now the proprietors seem to have enlarged their ideas to provide 7 ft of depth from Morwellham to Nutstakes, sufficient for 130-ton craft, and a basin there. A meeting in June, soon after the passing of the Act, made a call on shares 'to commence immediately their operation in cutting the said canal, and improving the said navigation, which last can only be executed during the summer months'.[18] In August a contractor was sought to deepen the river for the 600 yards from Impham Quay to Nutstakes, build the basin there, and also a lock 80 ft × 20 ft,*[19] and work began.

* As advertised. As built it was 70 ft × 20 ft.

About this time the company seems also to have dredged a 5-ft channel as far as Blanchdown.

Presumably because they had not raised all their money, and the difficulty of the times, they now had it in mind for the time being to build the canal section only as far as Horsebridge, at a cost of about £18,000, and employed James Murray to resurvey this part. They then went back to Rennie for his opinion. On the river section he was critical of their decision to deepen to 7 ft and build a basin at Nutstakes, whereupon they seem to have cancelled its construction. On the canal section he proposed a top width of 20 ft and depth of 3 ft, to take 10–12 ton boats if locks were used, or 5–6 ton for inclined planes powered by water-wheels, on which boats fitted with wheels would run. There was not much difference in the cost, but he preferred locks each holding up to four boats. He was sure that the craft used must be small, because 'should ever the Canal be continued to Bude, which I consider the main Object', boats must be able to pass the planes that would inevitably be necessary on the summit.[20]

However, the canal section was never started, and for practical purposes the navigation ended at Newbridge. In 1801 the company, who by now had built a quay there beside the Tavistock–Callington road, advertised it to let. They described it as 600 ft long by 40–60 ft broad, the upper part enclosed and suitable for landing coal, deals, etc., and for shipping corn and slate, the lower part not yet enclosed, but useful for laying timber, ore, culm and limestone. They offered to build limekilns and warehouses if required, and added:

> This quay has been lately erected by the proprietors of the Tamar Manure Navigation, very commodiously situated for . . . goods passing through the Tamar Canal in barges to and from Plymouth, which may be conveyed from the New Bridge to Milton, Lifton, Launceston, etc.[21]

Apparently it was not taken, for the company then put in a wharfinger, William Dugdale, built a three-story granary, two kilns and stabling, and in 1802 ran a 60 ton barge thence to Plymouth, advertising that as the aim of the undertaking was public benefit, the proprietors would only charge reasonable dues in order to establish a free quay.[22] In 1810 the company gave notice of a bill to extend the river navigation upwards to Inny Foot to join their authorized canal line at Dunterton, but took the plan no further.

In all, the company seem to have spent about £11,000 on their works. They remained in existence throughout the century and beyond, but no further extensions were made to the navigation. In 1905, with a paid-up capital now of only £323, the company were still paying small dividends on this amount. Traffic in that year was 7,940 tons, of which one-third was coal, the rest bricks and granite. After the middle twenties coal for the gasworks seems to have transferred to rail and lorry, and about 1929 the river ceased to be navigable. The company went into liquidation during the second world war.

Tavistock Canal

Morwellham quay on the Tamar had been the medieval port of Tavistock. By the early nineteenth century craft of 200 tons were coming up to it to bring Tavistock the goods it needed and to carry away the produce of the copper mines. Copper prices were high because of the war, and new mines were being sought to succeed the successful Wheal Friendship, opened in 1796 or 1797. In these circumstances, and the early plan of the Tamar Manure Navigation promoters for a branch to Tavistock having been dropped, it was proposed at a meeting in Tavistock in March 1803, following a survey in 1802, to build a tub-boat canal 16 ft wide and 3 ft deep from the River Tavy at Tavistock to the Tamar at Morwellham, with a branch to the slate quarries at Millhill, the purposes being the carriage of copper ore from Wheal Friendship and the new Wheal Crowndale near Tavistock, slate and limestone from local quarries, goods to and from Tavistock, the discovery and drainage of new lodes, and the provision of water power. The cost would be £40,000 upon the estimate of Messrs Taylor & Hitchins, and the Duke of Bedford, who owned all the land through which the canal would be cut, and took dues for the use of Morwellham quay, gave the project his blessing.

The principal features of this short canal with a main line of about 4 miles, were the aqueduct over the River Lumburn, the 2,540-yd tunnel through Morwelldown, and the inclined plane with a fall of 237 ft, the greatest in southern England, that carried goods in trucks from the end of the waterway down to the quay. The canal was built with a current downwards from Tavistock, so that the water could drive mining and mill machinery in its course as well as help the greater number of laden boats.

Besides deciding to apply for an Act, this first meeting resolved That the duke of Bedford be requested to make a grant of a Mining Set, for working all Lodes discovered in the Course of the Canal, Embankment, Tunnel, and Collateral Branch, that the Proprietors of the Canal may have the right to dig for Copper, Tin, Lead, and all Metals and Minerals, to the extent of 500 Fathoms East, and the same distance West, of the Canal, upon each and every Lode discovered (providing His Grace's Lands shall so far extend) paying one-tenth part of the produce to His Grace, and to have the same set for the term of forty-two years.[23]

The estimate included money for testing any lodes that might be found. The Duke granted the mining rights, and also gave the land for the canal and took one-eighth of the shares.

The Act[24] was obtained in the same year of 1803, and work began on the tunnel on 23 August, the engineer in charge being the 24-year-old John Taylor, the manager of Wheal Friendship. Almost immediately copper was struck near the Tavistock entrance of the tunnel, and the find was developed as Wheal Crebor. The mine had an entrance adjoining the tunnel opening, its machinery being driven by a water-wheel from the canal's current. Separate superintendents were appointed for the canal and the mining work, and the capitals were soon separated. The same committee seems to have managed both concerns till 1812, when, at a meeting of Adventurers in the Mines belonging to the Tavistock Canal, a separate committee was appointed. This was after further discoveries had been made, recorded as follows:

Great discoveries have recently been made in the tunnel under Morweldown, which forms part of this canal, rich veins of copper ore of amazing thickness begin to shew themselves, and promise an abundant harvest of profit to the proprietors of this spirited undertaking.[25]

In 1811 Wheal Crebor produced 1,308 tons of copper, and thenceforward to 1820 2,000 or 3,000 tons annually, the profit in 1813 being £5,462. From 1821 to 1856 only 6,269 tons were mined, and by the time a revival came from 1860 to 1894, when 32,306 tons were produced, the canal's best days were over.[26]

The cutting of the unlined tunnel went on steadily, the men suffering from bad air and an inflow of water that had to be pumped away, while the canal from the tunnel to Tavistock was built. By the end of 1809 the canal between these points was finished, and about half the tunnel. This was so small, about 12 ft in height from roof to invert, and 7 ft wide, that not many men could work at once, even though shafts had been sunk from the top of Morwelldown, and therefore progress was slow. At this time 30 or 40 men were working in it. On 21 August 1816 the two ends were joined, and it was reported that it would be ready

XI. (*above*) Share certificate of the Exeter & Crediton Navigation; (*below*) the Cann Quarry Canal about 1956; on the left, the towpath and site of the later tramroad, with rails still in place

XII. Tavistock Canal: (*above*) A section in rock cutting; (*below*) the north portal of Morwelldown tunnel, dated 1803

for boats early in the following year. The report for 1816 goes on:

To complete the communication with the Quays at Morwelham a piece of open cutting has been made, from the south end of the Tunnel to the side of the hill above the river; and after duly weighing the merits of various plans for transferring from the Wharfs to the Canal level, the Committee have adopted that of an Inclined plane, furnished with iron railways, on which carriages fitted to transport boxes which may contain ores, coal, lime, etc., are made to ascend and descend by the application of a machine driven by water supplied from the Canal.

The simplicity of this mode is a great recommendation of it; the whole is in a state of much forwardness, and is executing by competent workmen from a model which has been approved by the Committee and their Agents.

This description, if read with that given by R. Hansford Worth,[27] who himself inspected the remains of the plane in 1887, and by von Oeynhausen and von Dechen in 1826–7,[28] makes it clear that boats were not carried up and down the plane, though there are at least two statements to the contrary.[29] The description I have quoted mentions boxes for the transhipment of goods, but as the waggons on the plane were made to tip, I fancy they were in the end not used. There were two lines of rails, at first of channel type, and later, because these were badly cut by the flanges of the truck wheels, flat-headed. There seems to have been more than one truck on each line of rails, for Worth speaks of 'the ascending train of trucks'.* The plane varied in inclination, but reached 1 in 6, and the trucks were built with one pair of wheels of 4 in greater diameter than the other, in order to keep them level. The incline was worked by a water-wheel driven by waste water from the canal, and Worth describes it as follows:

The wheel itself . . . is about three feet breast, and twenty-five to thirty feet in diameter. On its axle is fitted a somewhat ponderous bevelled cog-wheel, into which gears another of similar dimensions. This is fixed on a stout wooden axle, which rises almost vertically, and moves, by the aid of another pair of bevel wheels, another horizontal wooden axle about one foot eight inches square in cross-section. On this last is fixed a large drum, on to which winds the chain, of four-inch link and three-quarter-inch iron, which was used to draw up the ascending train of trucks laden with coal, lime, etc. Another large drum, with a wire rope wound on it, is so connected with this last by

* This may be confirmed by the quotation on p. 133.

I

a pair of ordinary toothed wheels that the descending train, laden with copper ore, was made to assist the water-wheel in drawing up the ascending trucks.

The canal had taken thirteen years to build, and meanwhile war and boom had given way to peace and slump. Tavistock had so far been prosperous by reason of its mines, the population having risen from 3,420 in 1801 towards the 5,483 it reached in 1821. From the beginning of the peace, however, the mining industry continued only moderately till 1845, when there was another revival till about 1870, during which time the rich Devon Great Consols mine was worked, and the population increased again from 6,272 in 1841 to 8,965 in 1861. The committee, sadly and almost with premonition, reported in September 1816

That the Tonnage will fall very short of the original estimate cannot be doubted; at the time when this was made the Mines in the neighbourhood were in their most flourishing condition, or rapidly advancing to it. Agriculture encouraged by high prices was improving everywhere, and the use of lime, so essential to the good cultivation of the soil of the district through which the Canal passes, was very great and increasing. Slate quarries, then making large returns, were in full activity, and new ones opening. In all these sources for an oecumenical [*sic*] mode of carriage, the greatest reverses have taken place, and at present a stagnation of enterprise and consequent inactivity prevails.[30]

The opening ceremony was, however, carried off with aplomb and a good deal of courage, given the length and smallness of the tunnel: on 24 June 1817,

At eight o'clock between 300 and 400 persons of all ranks embarked in nine boats constructed of sheet iron* ... the whole party proceeded in their aquatic subterranean excursion with the greatest order and regularity under a salute of banners bearing appropriate inscriptions together with a party of miners and others dressed uniformly with ribbons in their hats having on them the words 'Success to the Tavistock Canal'.

More than 300 of the party kept their nerve, and entered the monument of industry and perseverance, with rather awful and somewhat sublime sensations. The timidity of the ladies was, however, soon relieved by the reverberating sound of music

* The following extract from the *Exeter Flying Post* of 4 April 1811, is of interest: 'There is now building at the Mount Foundry Iron Works in Tavistock an entire iron boat, which will carry eight tons; it is nearly finished, and will be launched on the Tavistock Canal in the morning of Easter Monday.' The first iron barge was, however, launched on the River Severn in 1787.

from the band of local performers in the several boats, which
contributed much to dispel the gloom and so lessed the appre-
hension at a depth of 450 feet from the summit of the hill. . . .
The pleasurable sensations excited by the approach to daylight
on again enjoying open sunshine were indescribable.[31]
A crowd of 5,000 people waited at the other end of the tunnel, and
were amused by the precautions the voyagers had taken to protect
themselves from drips from the tunnel roof. A salute of 21 guns
was fired as the boats moored, the working of the inclined plane
was demonstrated, and then there were refreshments, toasts, and
dancing.

The cost of the canal had been £62,000, the excess over the
authorized capital of £40,000 having been raised by calling £155
on each £100 share and not by mortgage. The proprietors cour-
ageously decided to proceed at once with the Millhill branch to a
slate quarry, and this was finished before the end of January 1819.
It was 2 miles long and cost about £8,000. The branch rose by
19½ ft to Millhill, with an inclined plane instead of locks because
water was scarce on the upper level. von Oeynhausen and von
Dechen's full text make its actuality clear. It was double-tracked,
with less of a gradient than usual. Two cradles were counterbalanced, the loaded boats going down pulling the empty ones up
with the help of three horses.

Trade on the canal was so bad that it was decided to charge tolls
a good deal less than those authorized: 1s per quarter on coal
against 2s, and 1s per ton on slates against 2s. To these charges
must be added the freight costs and the Duke's tolls as owner of
Morwellham quay. The canal then settled down to carry between
15,000 and 20,000 tons a year in its small boats, 30 ft by 5 ft, and
to earn an average profit of £600, the tolls being as follows:

Years	Average £
1821–25	1,083
1826–30	1,032
1831–35	778
1836–40	1,071
1841–45	840
1846–50	743

They then remained at about £750 a year till the railway opened in
1859.

A maiden dividend of £2 per £155 share was paid for 1820, and
the highest among those recorded is £5 per share for 1827, for
which there was a special reason. On 15 January 1828, because of

the difficult position of the mining part of the concern, it was resolved:

That it appears to this Meeting desirable to alter the Plan on which the Mining Operations have been carried on by the Canal Committee,* and that in future the Mines be carried on by such only of the Proprietors as evince their desire of continuing Mining Operations, and paying any Calls which it may be found necessary to make; and that the Income arising from the Navigation shall be annually divided among the Proprietors.

From this report it appears that since 1820 part of the profits from the canal had been put into the mines; the position was therefore regularized by paying a dividend of £5 on the canal shares and making a call of £5 on the mining shares. The usual dividend did not, however, exceed £3 per share, or just under 2 per cent. In 1829 the mining company was wound up and transferred to a new concern.

There was a heavy fall in tonnage in 1831, apparently due to the closing of mines, and in the following year tolls were reduced on condition that the carriers also reduced their charges. This policy was successful, and a steady improvement took place from 15,137 tons and tolls of £740 in 1833 to 20,009 tons and £1,167 in 1837. In the latter year, however, the company restored the tolls, and by 1844 the tonnage was down to 13,366, and the tolls to £763. The company now let the canal to Messrs Gill & Rundle the carriers at £700 a year and an extra payment if tonnage increased beyond an agreed figure, but there was no increase, and the agreement was not continued after 1853.

About the beginning of 1844 new slate quarries were opened at Millhill, and the owners agreed to send all their output to the Tamar for ten years at 2s per ton if the company would reopen the Millhill branch, which appears to have been closed some time after 1831, or lay a railroad. In July the company agreed to build a railroad at a cost of £1,200. It was opened at the end of the year, the actual cost being £1,381. However, the quarries were not a success, only 744 tons of slate being carried in 1848, 440 tons in 1849, and 676 tons in 1850. Meanwhile the tolls on the canal had been again heavily reduced in 1845, and this policy succeeded in raising the tonnage carried and keeping the revenue steady.

About this time the production of ore from the Devon Great Consols mine began to increase steadily till the middle sixties. No traffic was brought to the canal, but the wharf space at Morwell-

* It seems that though there were nominally two committees, a canal and a mining committee, in fact they consisted of the same persons.

ham had to be increased so that a number of vessels could be loaded at the same time, and a horse railway was built to replace the road from the mine. Morwellham was now

a scene of busy industry, with its unloading barges, and shouting sailors, and hammering workmen, and train of waggons ascending or descending the inclined plane. A quantity of ore is here shipped off to distant smelting-houses. It is curious to enter the well-swept yard, and observe the different wooden shafts down which distinct ores from various mines are poured. Then it is to be collected, and placed on board the vessels bound for distant quarters. These ships in return bring coals, and limestone, and many other commodities.[32]

At the end of the decade, in 1850, the goods being carried on the canal were:

	Tons
Sundries	7,548
Limestone	3,130
Copper Ore	2,499
Slate	676
Granite	83
Mundic (pyrites)	94
	14,030

In 1853 a curious state of things came about, resulting from a condition in the Act that no member of the managing committee should live more than 5 miles from Tavistock, except the Duke of Bedford and his principal agent in London. In this year only one member fulfilled the condition, and there was no one else qualified and willing to act. For the rest of the life of the canal its affairs were managed without strict authority by two or three of the principal proprietors, who called themselves the committee of management, and who met fairly regularly, though with frequent adjournments for lack of a quorum.

The following year saw the first threats of the railway competition between Tavistock and Plymouth that was to become actual in 1859. They led the committee to renew the machinery of the plane at a cost of £300, to set on foot inquiries about the cost and possibilities of a tug, and in 1859

... with a view to reduce the expence of Traction, it is resolved to accept Mr. T. Knight's Tender for propelling the Boats thro' the Tunnel by means of 2 Water Wheels and wire rope, &c., to be rendered complete in all respects as regards Wheels, Hatches, Wheelpits, and Watercourses &c and to the entire satisfaction of the Proprietors for the sum of Two Hundred Pounds, the wire rope to be paid for by the Canal Proprietors and the Clerk to be authorized to order the galvanized wire Rope direct or thro' Mr. Knight.[33]

In spite of alterations to the tunnel the plan was not a success, and in the following year

It having been found that the wire rope, on account of its friction with the sides and bottom of the Tunnel, does not answer the purpose of traction required of it, resolved that the Secretary be directed to write to Messrs. Morton & Co enquiring what amount they would be disposed to allow for the rope on its being returned.[34]

The disposal of the rope occupied the intermittent attention of the committee for the rest of the canal's life.

The South Devon & Tavistock Railway, later extended to Launceston, reached Tavistock from near Plymouth in June 1859, and in the following year the canal tolls were reduced to meet the new competition. On 10 September 1866 the committee recorded:

That a very considerable reduction having taken place in the traffic and dues of the Tavistock Canal Company since the opening of the Tavistock and Launceston branch of the South Devon Railway it was resolved that His Grace the Duke of Bedford be applied to to kindly assist the Tavistock Canal Company to compete with the Railway Company by reducing the Canal Dock dues at Morwellham.[35]

The Duke seems to have been evasive.

The Millhill branch, a tramway except for the short remains of the original waterway as far as the Tavistock–Gunnislake road, had once again been closed for some time, and attempts were made without success to sell the rails to the Lee Moor company. The company then considered whether they should dispose of the land on which the branch stood to the Duke, and at a meeting on 17 May 1870 the idea was extended to the whole canal:

With relation to the present position of the affairs of the said Canal Company it appears to the Proprietors present that it would be desirable that if it could be legally effected an arrangement should be made with the Duke of Bedford for conceding the Canal with its properties and effects to His Grace upon fair and equitable terms and with such a view that a committee be appointed to confer with His Grace's agents upon the subject.[36]

The members present agreed to offer their shares to the Duke at £10 each, and recommended the others to do the same. The Duke, however, declined, and in August an unsuccessful attempt was made to farm the tolls. A long legal argument then ensued upon whether the Duke had the right to resume possession of the Millhill branch without payment when it had been disused for five years. It was still proceeding when at a meeting on 2 September 1872 the Duke offered to buy all the shares at £8 per share, and to pay half the cost of an Act to transfer the canal to him. A special meeting on 23 September agreed to accept his terms, and in May 1873 an Act was passed authorizing the sale for £3,200, there being no obligation on the Duke to maintain the canal.

I do not know what now happened to the canal. In 1883 it was included in the Canal Returns as still open, but by 1898 it had disappeared from the list. Some sixty years after the purchase Act a new use was found for the waterway. In 1933–4 the bed was cleaned out, a new cut made from the south end of the tunnel to a reservoir, and the water led thence by a pipe down the cliff to Morwellham, where the canal's current began to provide supplies to a hydro-electric plant. Like the Bude Canal, therefore, the old waterway has found a new and useful purpose.[37]

Torrington or Rolle Canal

Before the canal period coal and limestone had been brought up the River Torridge past Bideford to the tidal limit at Weare Giffard, where it was burnt in the great kilns that still stand there before being carried inland for farm use first by pack-horse and then by waggon.[38] Then in November 1793 a meeting was called at Torrington under the chairmanship of Denys Rolle—the Rolles were great landowners—'to take into consideration the propriety of carrying a canal from some navigable part of the Torridge above Bideford Bridge to Torrington, and to communicate with some or one of the intended canals in this county'.[39] Soon afterwards the idea had grown to a canal from Bideford to Okehampton, with a branch to Torrington, to be linked with another, the Public Devonshire, onwards to Exeter, and perhaps also with one to Bude, for which George Bentley as engineer, with Mr Tozer as surveyor, did a plan and estimate.

Denys Rolle, like Lord Stanhope on the Bude Canal, was a man who saw the advantages of using tub-boat canals with inclined planes in the west country. His idea in using this kind of construction was to avoid building canals through the fertile river valleys of the Taw and the Torridge, placing them instead high up on the hills where land was cheap, and the sea-sand needed. Some years before 1793 he had built a half-mile long canal to bring water to a sawmill, and had with his coachman and bailiff himself dug experimental sections to judge a fair day's work. The following extract from a letter of Denys Rolle's, dated 4 September 1793, shows that at that time a Bude Canal scheme to Okehampton* was linked in his mind with the Public Devonshire idea of a canal from Exeter to Bideford via Okehampton:

Lord Stanhope I have just new information of is at his manor of Holdsworthy levelling himself and forming Boats etc. for the projected Canal to cross the *Poor Lands* of Cornwall and Devon from Bude Bay to the Tamar next Plimouth and the Exe at Topsham & the great Canal to Taunton & Bristol. The former's first meeting I attended and a Gentleman writes me some of my intimations were somewhat approv'd as also at Okehampton, respecting the carrying through the Poor Moor Lands from the rise of the Taw to near my sample of Canal & then to divide into two Branches one to Barnstaple the other to Bideford—saving the Rich Lands from being cut through in the Vale of the two Rivers leading to those 2 Towns—and the Great Object of conveying Sea Sand, Limestone and Manure ... on the Poor Lands wanting it ... it saves a vast deal of Carting and Horse Carriage.[40]

The war then postponed these schemes, but in 1810 James Green, probably working for Rolle, prepared a plan for a Torridge Canal from Torrington along the eastern side of the river past Weare Giffard to enter it just above Hallspill opposite Landcross.[41] A Parliamentary notice appeared, but it came to nothing.

Canal schemes between Torrington and Bideford looked especially attractive because of the hilly and difficult road that then connected the two places. In early 1823 a plan was being discussed for the road we now have beside the Torridge; it would be shorter and easier.[42] Notice of a bill for this road from the Torrington end to meet one to be built towards it by the Bideford Trust was given in December. In March 1824 tenders were being sought for the Bideford end, and in August a number of the road's supporters dined together, after Lord Rolle had sent them a haunch of venison 'in testimony of his approbation of the intended improvements,'[43] which in the circumstances was handsome of him, as he had started building a canal.

After some discussion upon the merits of a railway, and a survey for one by Roger Hopkins, engineer of the Plymouth & Dartmoor

* An extension of the Bude Canal scheme to Okehampton was probably being investigated by the Nuttalls, who were at work when this letter was written. See p. 145.

tramroad and later of the Bodmin & Wadebridge Railway, Lord
Rolle in 1823 began the canal at his own expense and without
parliamentary authority, with James Green as his engineer. John,
Lord Rolle, was the son of Denys Rolle, who had died in 1797.
He had been a Member of the Parliaments of 1780, 1784 and 1790;
then in 1796 the peerage formerly held by his uncle, who had died
in 1750 without issue, was revived in his favour. He lived at
Stevenstone House near Torrington, and was a considerable
Devon figure.

The canal began at a river-lock and basin joining the River
Torridge below Weare Giffard, and about 1¾ miles above Bideford
bridge. After a stretch of nearly a mile beside the river on its
western side, the canal rose by a single inclined plane to its summit level. The plane was double-track, powered by a water-wheel,
and took small tub-boats almost certainly fitted with wheels, as
on the Bude Canal. It then followed the curve of the river to
Beam, where a very fine classical stone aqueduct of five round-
headed arches* carried the waterway across the Torridge, the
other bank of which was then followed to the limekilns at Taddi-
port, warehouses, offices and the then existing Canal Tavern at the
bottom of Mill Street, Great Torrington, and on to end at New
Manor Mill, a mile above the town. The total length was about
6 miles.

This mill, in a crenellated style that must have gone rather oddly
with the canal, was built beside the basin by Lord Rolle to replace
the old manor flour mill near Mill Street. A new road, and a bridge
over both the canal and the river alongside the new mill, were
later built, the bridge being opened in 1843, the year after Lord
Rolle's death.

The main purpose of the canal was to carry limestone from
south Wales, especially Caldy island; it was unloaded at Appledore
or Bideford into barges to be taken to the river-lock, where it was
again transferred to tub-boats, usually worked in trains, to be
carried to Lord Rolle's kilns at Torrington. Coal was also imported in the same way, and agricultural produce sent away.

On 11 August 1824, when Lord Rolle visited Torrington: 'On
his entry, his Lordship was met by the Mayor, Corporation,
Feoffees, and a great number of Inhabitants, who escorted him to
the Town-Hall, to a public breakfast.' Afterwards he laid the first
stone of the great aqueduct, made an appropriate speech amid loud
cheers, and then sat down to an elegant dinner, while beer and
cider were issued to the crowd, 'who amused themselves in rural

* Now used for the driveway to Beam House.

diversions'.[44] His Lordship seems to have been of a military turn of mind (he was a colonel of the South Devon Militia and of the Royal Devon Yeomanry), for in addition to the crenellated mill, it is recorded that at the foundation-stone-laying ceremony John Hopgood was wounded in the arm by the bursting of a cannon placed on Furzebeam hill. The sum of £12 2s 6d was paid to the surgeon who attended him, and he himself received one year's allowance. He was presumably one of Lord Rolle's men.

The new road was opened in mid-1826, the canal not until February 1827.

The first vessel to enter the works will be the *May*, of Bristol, 70 tons burthen, from Newport, laden with coals, as a gift of the Right Hon. Lord Rolle to the poor of Torrington and its vicinity. The *May* is provided with a striking mast in order to pass Bideford Bridge, from thence she will proceed to the entrance lock, and having entered it, her cargo will be discharged into four-ton boats—these will precede a number of similar boats laden with various articles up the inclined plane, past the beautiful and substantial stone aqueduct across the Torridge, and proceed to the town of Torrington.[45]

It was not until July 1829, however, that a gathering of a hundred people dined at Beaford and presented an address to Lord Rolle to celebrate the completion of both canal and road.

Before the canal had opened, a shipyard had been established by the entrance lock:

The Rolle Canal Company owned the premises and vessels built there were for their own use. The first vessel launched by the Company was named Lord Rolle, a schooner burthen 98 tons. The vessels brought coal and other goods from Wales and elsewhere to Bideford and Appledore. . . . For the New-foundland trade, three-masted schooners were built, while vessels were also built for trading purposes in the Mediterranean. The *Sedwell Jane* (1868), a three-masted schooner, 200 tons burthen, was the largest ship built by Richard Pickard and John Leonard, shipbuilders, at Sea Lock. The shipyard was closed in June, 1870.[46]

In 1831 a proposal, followed by a survey and estimate, was made for a locomotive railway to join the end of the Rolle Canal to Okehampton, so providing the port of Bideford with access to the interior of Devon, and extending the market for locally imported limestone and coal. There was a separate report upon a possible extension to Bideford, which would have saved the transhipment of goods at the river lock. No action followed. The idea was revived in 1845 as the Bideford & Tavistock Railway, and in 1852 as the Bideford & Okehampton again, to run down the Ockment valley to Torrington, and then to Bideford either using the Torrington Canal's course below Weare Giffard, or on a different line. The engineer was Roger Hopkins's son. It was not until 1925, however, that the line from Torrington to Halwill Junction was actually opened.[47] A lighter moment was provided by a correspondent of the *North Devon Herald* of 9 March 1871 who proposed a single-rail double mule-power tramway.

The cost of the canal is given by different sources, none of them original, at different figures; between £40,000 and £45,000 would probably be correct. As a result of a vestry meeting at Torrington in January 1833, which sought to re-establish the common rights of the inhabitants, an Act was passed in 1835 to enable Lord Rolle to buy the common rights of the parts of the Torrington commons which had been used for the canal works, for a sum which was afterwards assessed at £150. The Act also contained the curious provision that the owner of the canal should not sell by retail anything except coal, culm, lime, flour, and timber within the parish of Great Torrington, an indication of the trading business that Lord Rolle had built round the canal.

At some time before September 1852, Mr George Braginton, son of Mr Richard Braginton who at that date had been steward and manager of the Rolle estates for nearly forty years, leased the Torrington Canal.* He was a considerable local figure, several times Mayor of Torrington, a merchant, and the proprietor of a local bank. In September 1856 a local newspaper recorded:

> On September 13th., Mr. George Braginton, banker, Mayor of Torrington, entertained a large number of friends, and also the numerous workmen employed on the Rolle's Canal, in honour of the fiftieth anniversary of the marriage of his parents, Mr. and Mrs. Richard Braginton. The family feast was held at Mr. George Braginton's, Moorhouse, St Giles, and the supper to the workmen at the Rolle's Arms, Torrington. At the supper Mr Robert Brinsmead, the oldest hand on the canal, one who helped to cut the first sod, and manager of the canal since its opening, should have presided, but in his absence through illness, the chair was taken by Mr. John Sandford, clerk to the bank.[48]

I do not know for how long Mr Braginton was lessee, but in 1865

---

* Lord Rolle had died in 1842, leaving his property to the second son of Lord Clinton, the Hon. Mark Trefusis, on condition that he changed his name to Rolle. He was the owner till the canal closed.

his bank failed, and he was made bankrupt. Then or earlier possession of the canal reverted to the Hon. Mark Rolle.

In 1845 plans had been deposited to continue the railway from Barnstaple to Torrington, but the line was not proceeded with. In February 1853, however, it was stated at the half-yearly meeting of the North Devon Railway Company that a new concern, the Bideford Extension Railway Company, had been formed to build the line from the company's docks at Fremington to Bideford. George Braginton, then the canal's lessee, was chairman of the board of the Extension Company whose line when built might be expected to increase the trade on the Torrington Canal. The first sod of the railway was cut in August; in July 1854 the North Devon's own line to Barnstaple was opened, and in October the extension to Bideford. A year later the local newspaper said that railway traffic to Bideford greatly exceeded expectations.

Meanwhile, in March 1854, a public meeting had been held at Torrington to advocate the extension of the Bideford Railway to that town, and at the opening of the line the Mayor of Torrington said he hoped for such an extension. Nothing occurred to interfere with the canal, however, except an improvement in the local roads, till a public meeting held in Bideford in February 1865 to consider the best way to get direct railway communication with Torrington, Holsworthy, Launceston, Plymouth, and the west of Cornwall. An instalment of these demands was met when in August the South Western, which had taken over the North Devon and Bideford Railways, announced a forthcoming bill to build a line to Torrington. The Act was passed, but the cost of construction set against the probable receipts deterred the company. Then 'under pressure from the shareholders, the London and South Western Railway Company went to Parliament for powers to abandon it; but were firmly withstood by the Hon. Mark Rolle, thanks to whose gallant opposition permission to violate the engagement to construct the line was refused'.[49] The spectacle of a canal owner leading the demand for a railway against a railway company anxious not to compete with his canal is probably unique. A railway representative at the opening of the line said of this situation:

> The question was how far Mr Rolle, whose interests were identified with the canal, would be prepared to meet them. . . . Mr Rolle met them in the most liberal spirit; he would be glad, he said, to see the canal superseded by a more useful and convenient mode of transit and communication.[50]

It was agreed with Mr Rolle that the canal should close, and that part of its bed should be used for the railway. Work was begun in May 1870, and the line was opened on 18 July 1872, at a cost of almost £100,000. I do not know the exact closing date of the canal, but it was probably during 1871. The upper part, from Torrington station to New Manor mill, was turned into a toll-road called the Rolle or New Road (this was being built in April 1875) but it was not a success. The road was later offered to the town council, and then abandoned. The channel of the aqueduct at Beam became a private drive, the railway cut across the centre of the inclined plane, and the rest of the line became derelict.

The Braunton Canal and River Taw schemes

In 1810 a parliamentary notice for a bill to authorize land drain-age and enclosure works at Braunton and Heaton Punchardon included a navigable canal 'from the lower end of Wrasenton Marsh . . . to certain lime kilns, and other places at the higher end of Vellaton, and thenceforward into Braunton Field . . . and other-wise as far as may be necessary'. An Act was passed in 1811, but without the canal scheme.

In 1813 James Green suggested that two drainage channels made in Braunton Marsh should be connected by a lock

to give a passage by boats into the interior of the north or high part of the . . . Marsh, and in 1821 deposited a plan for what seems to be a combined drainage and navigation canal from Broad Pill on the Taw, where there would have been a lock, to North Sluice on Braunton embankment.[51]

It was not built. Small craft were, however, accustomed to pass through the sluice at Pill's Mouth on the tide with cargoes of gravel, limestone and coal, and work up to Vellator below Braun-ton. About 1850 a local landowner, Sir Frederick Williams, altered the course of the united streams of the Caen river and the Knowle Water, which ran from Vellator to the Taw, in order to enclose Horsey Island. A new cut was made east of the island, and at the same time a quay was built at Vellator big enough for ships of 130 tons, at which charges were now levied. Some gravel is still carried up this new cut to Vellator.[52]

In 1845 a plan was made for a very big canal from Barnstaple up the Taw as far as Umberleigh bridge, which one would have taken for a drainage channel were it not that a lock is shown at the lower end.[53] It was never built. Barges and small boats could, however, get up to Bishop's Tawton and New bridge on a flood-tide.

CHAPTER X

The Bude Canal

✦✦✦✦✦✦✦✦✦✦✦✦✦✦✦✦✦✦✦✦✦✦✦✦✦✦✦✦✦✦✦✦◆✦✦✦✦✦✦✦✦✦✦✦✦✦✦✦✦✦✦✦✦✦✦✦✦✦✦✦✦✦✦✦✦

SHELLY sand, carried inland from many Cornish and Devon beaches, was in great demand as a soil conditioner, its calcium carbonate content making it an alternative to lime. Moore, in his *History of Devonshire*, writes of Bude:

The harbour is dry at low water, and its bed is formed of a fine bright yellow sand composed almost wholly of shells. This sand, from time immemorial, has been used as a manure for the surrounding country. Before the canal was cut, the harbour presented a very animated appearance at every return of low water during the tilling season, being thronged with waggons, carts, pack-horses, and herds of mules and asses, loading and carrying off the sand into the interior of the country.*

The canal to which Moore refers is the Bude Canal. Something similar was first projected early in 1774 by John Edyvean, whose St Columb Canal (see p. 165) had been authorized the year before. It is likely that he had had a line surveyed by Edmund Leach and John Box before the first meeting was held at Launceston in February 1774, for they decided straightaway to seek a bill. A winding Tamar Canal was proposed, over 90 miles long and with a 68-mile summit level, to cover the 28 miles of direct distance between Bude and the River Tamar at Calstock, with five inclined planes. It was to be a small affair to carry 10-ton tub-boats, with a towpath on each side. The winding course was not seen as a disadvantage, for the line would run high up on the hillsides, where the ground needed improvement, and would make the necessary sea-sand easily available. Edyvean, giving evidence before the Lords' committee, spoke of the inclines as his own inven-

---

* How it was valued at the time is shown by G. B. Worgan's statement in *A General View of the Agriculture of Cornwall*, 1811, p. 126: 'Long experience has proved, that sea-sand is a fertilizer of the soil, as well as for pulse or roots, and excellent for pasture. It is of such estimation with farmers, as materially to affect the value of estates, according to their nearness to, or distance from, this manure. It is frequently carried fifteen miles inland.'

tion. They seem to have involved using trucks, and so the transhipping of cargoes at each plane. Edyvean hoped to sell sea-sand at Calstock at 2s 0d a ton.

Support was obtained from local people, notably Sir John Molesworth and the Call and Rodd families who were later concerned with the Bude Canal; that indefatigable but undoubtedly eccentric newspaper correspondent Sampson Grills wrote:

This fag-end of the kingdom will be visited with great pleasure, as a country very capable of improvement, and for which nature has done a great deal, but art, as yet, little or nothing—Tis to be hoped the Clergy will make no unreasonable demands through the improvement,[1]

and an Act[2] passed in 1774. The line was 95 miles long, rising to 300 ft, from Bude to Marhamchurch, Poundstock, Whitstone, North Tamerton, Boyton, Werrington, North Petherwin, Tremaine, Egloskerry, and by Launceston to Laneast, Lewannick, North Hill, Linkinhorne and Stoke Climsland to the Tamar at Calstock. A capital of £40,000 was authorized, with £20,000 more if necessary. Maximum tolls were very low: 2s 0d for coal, sand and lime, even if carried the whole length, and 3s 0d for merchandise. Commissioners were empowered to raise or lower tolls, having ensured that the shareholders got 6 per cent on their money, and exceptionally, the company was specifically forbidden to lease itself, or to be concerned in carrying upon its waterway.

A first call of £10 per £100 share was made in June 1774, and a meeting was called for December 'in order to consider of the effect of Mr. Edyvean's machine, and to make a further advance of money'.[3] By August 1775, however, discouragement is evident in a meeting notice 'in order to settle some decisive measures for either prosecuting or giving up the scheme'. Probably the low capital and toll authorizations, and the prospect of what amounted to a limited dividend, and maybe also Sir John Molesworth's death in October, discouraged a move. John Smeaton was called in to advise, and in his report of 1778, re-estimated the same line at £119,201, remarking that 'the county of Cornwall . . . seems but ill-adapted for the making of canals across the country, being so very frequently intersected with valleys, that to preserve a level for any considerable space between two given points, it becomes necessary to go through a vast meandering course'. He suggested as an alternative that the Bude river should be locked for 3½ miles; that 6 miles of canal with three planes should be built to the Tamar, and then 15½ miles of river navigation with ten locks to Greston bridge, whence a further stretch of locked river could carry the navigation to Calstock, or a branch canal could take it up to Launceston. His estimate to Greston bridge was £46,109. As a second best, he recommended a canal from Bude to Launceston only, but shortened to 34 miles, with five planes.

Edmund Leach, fascinated by the problems raised in his survey of the Tamar Canal and another between Liskeard and Looe, spent some years in working out in meticulous detail what might be done, publishing the results in his book, *A Treatise of Universal Inland Navigations*, about 1785. He re-estimated the original Tamar Canal line at £88,740, below Smeaton's figure but much higher than Edyvean's, and considered that 20 tons of sand could be delivered by canal to Launceston for £4 13s 2d against £12 5s 5d by land carriage. Alternatively, he suggested a line adapted from Smeaton, for a canal up the Bude valley, and thence by Tamerton bridge to Calstock, 40¾ miles, with a short branch. There were to be two planes and a long tunnel on the main line, and a third plane on the branch. He also suggested a longer branch to Launceston.

He also gave much thought to the inclined planes. He saw the expense and trouble involved in transhipment of cargoes at each plane, and worked out a method of avoiding this. His planes were not railed, but boarded, on which two cradles carrying the boats dry ran on rollers. Motive power was partly by water-wheel or walking-wheel (that is, worked by men walking within it, as some contemporary cranes were operated), but with part of the movement provided by running water into a tank beneath the uppermost cradle. He did not propose a pure counterbalanced plane, seemingly because he had not thought out a way of braking the descending cradle, and therefore had to keep control of the operation by means of a wheel. Leach's proposals have interest, as they preceded the first inclined plane to be built on an English waterway, that on the Ketley Canal in 1788; he seems not to have heard of earlier continental attempts, or of those on the Tyrone Canal in Ireland.

Maybe partly as a result of Leach's book, partly because of interest at the Tamar end in what became the Tamar Manure Navigation, a canal scheme was revived with the strong support of Lord Stanhope, who owned an estate at Holsworthy. This Lord Stanhope, born in 1755, the father of Lady Hester Stanhope, was a supporter of the French Revolution and of the reform movement at home, and was also a scientist and inventor of eminence, especially in marine engineering and printing. In 1790 and 1807, for instance, he took out patents for propelling vessels by steam.

As the result of a meeting in April 1793, to consider a canal from Bude Harbour past Holsworthy to Hatherleigh, supported by such local people as Sir William Molesworth, Sir John Call, Denys Rolle and the Cohams, at which £10,450 was subscribed by forty-two people, about the sum they thought it would cost, Bentley and Bolton made a preliminary survey, which showed the idea was practicable, but more expensive. In July they were commissioned to make a detailed examination, but being presumably too busy with other jobs, they were replaced by John and George Nuttall, who put an estimate to a meeting at Holsworthy on 25 October, with Lord Stanhope in the chair.

> Ever since our Arrival in Devonshire we have been unremittingly employed in surveying, and planning the Line of a Canal . . . we calculate, that upwards of 250,000 Acres will be benefitted by this Undertaking . . . we consider (it) as one of the safest and best that was ever proposed in this Kingdom.

This plan was for 75 miles of small canal, using boats of only 2 tons, and the proposed lines included those from Bude to Tamerton bridge and Blagdonmoor that were afterwards built on much the same routes. The Nuttalls, however, continued the Blagdonmoor line to Dunsland Moor at the head of the Claw valley and also proposed a long branch from near Holsworthy by way of Bradworthy to the River Weldon, and then to the River Torridge near Newton St Petrock. Thence the canal would run down the Torridge valley to near Hatherleigh.

The Nuttalls proposed that the long level pounds of their canal should be connected by three short railways of a total length of nearly 4 miles, and by one double-track inclined plane down to the Torridge valley. Their report says:

> The expence of loading and unloading the Boats, at the Junction of the Rail Roads, with the different Reaches of the Canal, will be saved, if Earl Stanhope's Plan for the Boats and Rail-Road Wheels be adopted.

What this idea was John Farey tells us:

> Earl Stanhope . . . proposed . . . to have iron rail-roads of gradual and easy ascent, on which boats of two tons were to be used, suspended between a pair of wheels of about 6 feet diameter, and to be drawn up or let down the same by a horse.[4]

The meeting decided to seek an Act, and thanked Lord Stanhope for his

> indefatigable Zeal and Perseverance in pursuing and personally investigating the best Means by which this great Undertaking, so very beneficial to the adjacent Country in particular, and to

K

the Improvement of Agriculture in Devon and Cornwall in general, might be accomplished, in spite of many natural Difficulties not common to other Counties.

Robert Fulton, twenty-seven years old, till then known as a painter, and not long arrived from America, had read a description of the proposed Bude Canal, and wrote to Lord Stanhope on 30 September 1793.[5] Fulton at this time had done no thinking about canals, and many of his subsequent ideas were probably stimulated by his correspondence with Lord Stanhope that lasted from 1793 to 1796.

Fulton's original letter had suggested that locks might be avoided by inclined planes worked by adding water to the descending weight, and also mentioned the possibility of steamboats. Lord Stanhope replied that his idea for working a plane had already been thought of by Edmund Leach, but that he was also working on steamboats, and was interested in Fulton's ideas. In 1794 Fulton took out a patent for improvements in inclined planes, in which wheeled boats (afterwards used on the Bude Canal) were mentioned, not, however, to run up the planes, but up small inclines in order to enter the caissons or tanks in which they were to float while being moved from one level to another. The patent also described a vertical canal lift. This idea seems also to have attracted Lord Stanhope in a different form; in 1796 he consulted Rennie and Fulton about what he called a pendenter. Fulton replied that he saw many objections to lifts, and advocated planes, which he was in favour of working by 'the tub passing through the pit', which seems to be a forecast of the tub-in-a-well system used later at the Hobbacott Down plane of the Bude Canal.

In October 1794 Fulton had first begun to be practically concerned with canals when, with a partner, he contracted to cut a section of the Peak Forest Canal. He then interested that company and its engineer, Benjamin Outram, in inclined planes to the extent of their sending a deputation to see those on the canals near Coalbrookdale, and of subsidizing of his book, *A Treatise on the Improvement of Canal Navigation* in 1796. In the summer of 1796 he surveyed for the proposed Helston Canal with its seven inclined planes (see p. 175), and in 1797 he left England. It is difficult to say whether he added anything to the ideas of Lord Stanhope; certainly he did little practical canal engineering in England.

It may be that Lord Stanhope's personal position as a supporter of reform, with the difficulty of the times, was the reason why the Bude Canal project did not then get started. Instead, action moved south, to make the Tamar navigable upwards to Tamerton bridge (the Nuttalls had provided for a connection by inclined plane were this to be done), and in 1796 an Act was obtained for the Tamar Manure Navigation (see p. 124).

Nearly twenty years later, in the interval between the end of the Napoleonic War and the Hundred Days, two gentlemen who possessed the Nuttall plans of 1793 discussed their revival, partly as a means of employing the poor. One of them went up to London to see whether Lord Stanhope was still interested in the idea, and would support it. He agreed to do so, but the revival of the war and then his death in 1816 postponed action. Not long afterwards, however, local support and interest having been widened, the new Lord Stanhope agreed to help, and in 1817 James Green and Thomas Shearn were invited to survey a line; they reported in the following year. Because this report determined the construction of the canal, it may be interesting to quote from it.

The committee prefaced the report by saying that, apart from sand, 'to procure coals is impossible, excepting at an enormous expence,' lime was needed if it could be supplied cheaply, and also 'merchandize and necessaries from Bristol, by way of Bude, which are now supplied by circuitous and expensive routes'. James Green wrote:

In consequence of the resolutions of your General Meeting, on the 13th of August last, directing me to examine the country between Bude Haven and Holsworthy, with a view to ascertain the practicability of forming a communication between the two places, by means of a Canal or Rail-Road, with a branch towards Tamarton Bridge; taking, as a basis, the lines proposed by the late Earl Stanhope,* and, as the leading object, the conveyance of sea sand into those districts which apply it as a manure. . . .

If . . . a Canal be resolved on, it ought to be on a small and inexpensive scale; and some of the same reasons which operate against a Canal on Lock principles, form serious objections to the adoption of a Rail-Road, inasmuch as 362 ft. of the whole elevation between Bude and Holsworthy must be attained before it can reach the line of the Tamar, and that within the space of five miles and a half. . . .

If a Canal, on which a horse may with ease convey 20 or 25 tons of cargo . . . can be executed for a sum not far exceeding that of a Rail-Road, I think it will be the most desirable way of effecting the internal communications so much wanted in these districts.

With this view of the subject, I have made my survey for a

* The 1793 proposal.

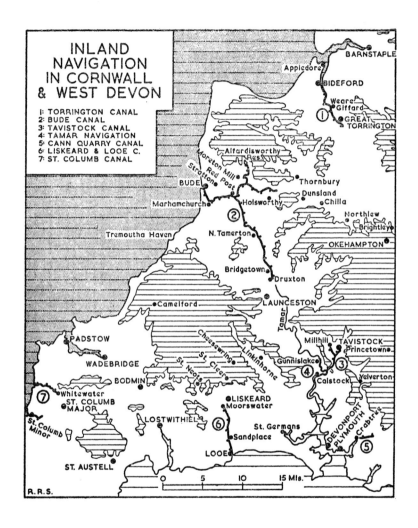

INLAND
NAVIGATION
IN CORNWALL
& WEST DEVON

1: TORRINGTON CANAL
2: BUDE CANAL
3: TAVISTOCK CANAL
4: TAMAR NAVIGATION
5: CANN QUARRY CANAL
6: LISKEARD & LOOE C.
7: ST. COLUMB CANAL

BARNSTAPLE
Appledore
BIDEFORD
Weare
Gifford
GREAT
TORRINGTON

Alfardisworthy
Res.
Moreton Mill
Red Post
Stratton
BUDE
Marhamchurch
Holsworthy
Thornbury
Dunsland
Chilla
Northlew
Brightley
OKEHAMPTON
Tremoutha Haven
N. Tamerton
Bridgetown
Druxton
Camelford
LAUNCESTON
Tamar
PADSTOW
Cheesewring
Linkinhorne
Millhill
TAVISTOCK
Princetown
Gunnislake
WADEBRIDGE
St. Cleer
St. Neots
BODMIN
Calstock
Yelverton
Whitewater
ST. COLUMB
MAJOR
LISKEARD
Moorswater
St. Columb
Minor
LOSTWITHIEL
St. Germans
Sandplace
DEVONPORT
PLYMOUTH
Crabtree
LOOE
ST. AUSTELL

0   5   10   15 Mls.

R.R.S.

Canal of small size, (viz) 10 feet wide at bottom, 3 feet deep in water, and 19 feet wide at top. I propose that the boats shall be constructed to carry 5 tons each, and that one horse shall draw at least 4 of such boats.

I propose to overcome the ascent from the sea to the summit-level, by means of inclined planes, up which these small boats may be readily passed by machines of a simple construction, moved by water. These machines may be executed at one-third the expense of locks and worked with one-third of the quantity of water which locks would require, with a saving of more than four-fifths of the time in passing them.[6]

He then goes on to propose a breakwater at Bude and a deepening of the harbour, to make it safer and to encourage a trade from Welsh and Bristol Channel ports that would benefit the canal. He also suggests that the first 2 miles from the harbour to the foot of Marhamchurch plane should be of larger size, 37 ft 6 in at top and 4 ft 6 in deep, so that 40-ton barges could take sand from the beach to be transferred at Marhamchurch to the tub-boats, and receive from them cargoes which could then be taken alongside vessels in the harbour. The two locks on this section would each be large enough to take one of these barges, or six tub-boats.

His proposed small-scale line from Marhamchurch ran to Red Post. Thence one line would cross the Tamar at Burmsdon (the Nuttalls had chosen the same spot) and run by way of Stanbury Cross (Holsworthy), Blagdonmoor, Dunsland Moor (the terminus of the Nuttalls's line), Chilla, Northlew, and Croft to Brightley bridge north of Okehampton; from this line there would be a feeder leading to the reservoir at Alfardisworthy, a branch to Virworthy, and also a branch to Thornbury from Stanbury Cross. From Red Post another line would run to Tamerton bridge. For these canals the estimate was £128,341, and the revenue was put at £15,083 a year, of which some £12,000 was to be derived from sand. Green also gave an estimate of cost (not exceeding £25,000) and of revenue for a possible extension from Tamerton bridge to Launceston, and suggested that such an extension might procure the completion of the Tamar Manure Navigation, and thus join the two Channels. It will be remembered that the Tamar Act authorized a line to Tamerton bridge itself, but evidently Launceston was now thought more likely if the scheme were revived.

The engineer finishes his report in fine style:

The time prescribed for my survey prevented my extending it further eastwards than Okehampton; but I am persuaded, that the summit-level may be carried many miles further in the direction of South Zeal, North Tawton, Bow and Crediton; by which route a junction may be easily effected with the proposed Crediton Canal,* and a direct communication opened from Bude to Exeter. Lines may also be extended from this summit-level to Bideford, Barnstaple, South Molton, and Tiverton; by which means, water communications with all parts of the County of Devon may be opened.

The committee added:

The nature of the soil in North Cornwall and North West Devon is such that from time immemorial resource has been had to sea sand as manure, and the received opinion is that the more sand is carried the greater the crop. Every good farmer exerts himself to bring home the greatest quantity of sea sand, and every landowner in these districts covenants with his tenant that a certain quantity shall be carried on every acre broken up for tillage.

Mr. George Call, later to be the company's chairman, estimated that 119,475 tons of sand would be carried annually over the canal, and a revenue of £14,733 would be derived from it.

An Act was obtained in the following year, 1819, and power given to raise £95,000, and a further £20,000 if necessary. The principal shareholders were: Sir William and Lady Call (£10,000) and George Call (£2,500); John Blackmore of Exeter (£5,000); Sir A. Molesworth (£5,000); Earl Stanhope (£4,000), and the engineer, James Green, backing his own scheme with £3,000. Green's proposals had been cut down to an estimated £90,623, probably to meet the capital available, and they were now for a barge canal from Bude to Hele (Marhamchurch), 2⅛ miles, and a tub-boat line thence to Red Post, 3⅝ miles beyond. Here one canal would run for 15⅝ miles to Holsworthy and on to Thornbury, with a branch to near Moreton Mill (1½ miles), that continued for another 2¼ miles as a feeder from Alfardisworthy reservoir, and another to Virworthy (3⅞ miles), also fed from the reservoir. The other line left Red Post by way of Tamerton bridge for Ridge-grove Mill near Launceston (19 miles). The final estimate was £91,617, of which £4,618 was for harbour construction. On the third reading of the bill, however, the Launceston line had been cut back to end about 3 miles short of that town, at Druxton (Cross-gate). An authorized branch of the Tamar Manure Navigation ran to Ridgegrove Mill, Launceston; it may have been opposition from this company to competition or a possible junction, were their line to be later built, that led to the Bude's being shortened.

* The Exeter & Crediton Navigation (see p. 116).

The proposed length of the Bude Canal was thus almost 46 miles.

Construction began on 23 July, when bells were rung, volunteers gave a *feu de joie*, and 12,000 people watched Earl Stanhope lay the first stone of the breakwater and cut the first sod of the canal.

> After the company had been regaled, dancing commenced in two long sets, which continued with the highest glee, gaiety, and hilarity for several hours. . . . Ten hogsheads of cider and many thousand cakes were given to the populace, and proper refreshments were supplied to everyone.[7]

John Kingdon was appointed 'Inspector' or resident engineer: extracts from his diary will give the feel of the work of construction:

Su 22 Oc 20	Remained at Bude in company with the Chairman* watching the effect of the High Tide and Seas on the Breakwater and Sea Lock. Counted on the work 340 men.
We 6 De 20	Placed twelve marks in the sand at Bude. . . . Party of Masons building a bridge near Hele Bridge where the course of the stream is finished. Parties at work on Abbacott† and Marham-Church incline plans and about 40 men excavating the Sea Lock Channel.
Th 14 De 20	The shaft at Marham-Church Incline for the Water Wheel to work in is fallen in. A gang of workmen dealing with it.
We 27 De 20	One of the whims at Abbacott† at work and one shaft expected to be clear of water this night. Plane partly formed and each shaft down about 70 feet and the adit driven about 60 feet.
We 24 Ja 21	25 men employed at the reservoir, some of whom are throwing back the bog and whallowing in rubbish. . . . Nearly the whole line from the reservoir to the Tamar is now open, but many parts in an unfinished state.
April 18	Employed with the chairman at Bude in getting mooring posts fixed for warping vessels from the sealock to the basin and fitting barge No. 1.
April 21	Took the barge No. 1 out of the Sea Lock and put on board her about 24 tons of sand. P.M. at tide time let her into the Basin, the barge drawing 3 ft 6 in. aft and 2 ft 10 in. forward.
April 25	Towed the barge to Hele Bridge. Discharged the sand on the wharf and moored her.

* G. C. Call.   † Hobbacott.

June 26   At the reservoir great leaks.
November 13   From Burmsdon to the end of the Lanson* line now
               open. 13 masons finishing the bridge at Burmsdon, 5
               men working on the quarry between Burmsdon and
               Anderton, no more from there to Red Post, the Gangs
               on the Lanson Line. I could not count the weather so
               bad they could not work.
December 22   Could not go on any part of the Line owing to a fall
               from my horse.

On Tuesday, 8 July 1823, the 'Bude Harbour and Canal, with
the inclined Planes and Railways' was opened. By this date the
large canal from Bude to Hele (Marhamchurch) had been built,
together with the breakwater at Bude; the line to Red Post; the
Thornbury branch as far as Blagdonmoor,† 2½ miles beyond Holsworthy, which had its own wharf at Stanbury Cross on the road
to Bideford, the Virworthy branch which left the main line at
Brendon Moor junction and was also the feeder from the reservoir
at Alfardisworthy, 70 acres in extent, with a capacity of some 195
million gallons, fed from the floodwaters of the Tamar, and the
Druxton branch as far as Tamerton bridge.

That respectable and liberal character, Mr. Blackmore, of
Exeter, had the gratification of seeing launched, on the water
of the Canal, two barges of thirty tons each, and many boats,
his property, and of receiving on board them manure and mer-
chandize for Holsworthy. . . . The Committee of Management,
supported by the neighbouring gentry, on the arrival of the
loaded boats at the point of debarkation, marched into the
town of Holsworthy in procession, the band playing 'See the
conquering hero comes', and hailed by the acclamations of the
populace of the surrounding country. . . . The dinner, provided
at the *Stanhope Arms*, was composed of the choicest viands; and
the hilarity, happiness and unanimity of all present, were most
auspicious.[8]
Among the many toasts were: 'Success to the Bude Harbour and
Canal, and to its speedy extension to Plymouth,' to Earl Stanhope,
the 'noble (though absent) Chairman General of the Committee,
and Friend to the Undertaking', and the 'Sire of the Canal', Mr
John Blackmore. It is notable that John Blackmore, here given
credit for having generated the project, was an Exeter man, and

---

* I assume Launcells.
† Some work was also done on the approach to a tunnel that was to have been
built a little east of Blagdonmoor, towards Thornbury.

that throughout the company's life its main office was at Exeter, shareholders' meetings being held there.

The first few weeks of trade in sea-sand were encouraging. By May 1824 the newspaper reported[9] that there were a hundred boats on the canal, trade was gradually increasing, and the Tamerton bridge–Druxton section was to be started. By July work had begun, and in August £50 shares, which at one time had fallen to £12, were selling for £30. To finance the extension and meet other liabilities, the company in 1824 borrowed £16,000 from the Exchequer Bill Loan Commissioners at 4 per cent, one-twentieth of the capital to be repaid annually. Two years later, unable yet to meet their liabilities, they borrowed another £4,000. Construction ended when the line from Tamerton bridge to Druxton (for Launceston) was opened in 1825. Neither the extension from Blagdonmoor to Thornbury, nor the Moreton Mill branch and feeder were built. The completed length was 35½ miles, to pay for which the sum of £92,367 was raised in 1,811 fully-paid £50 shares totalling £90,550, and £1,817 paid on those later forfeited or abandoned. Including the loan from the Exchequer Bill Loan Commissioners and a few thousands owed elsewhere, the cost of construction appears to have been about £10,000 less than the figure of £128,000 often given, which is probably taken from the figure of total expenditure in the balance sheet of 1828–9. This figure, however, includes not only the capital cost, but maintenance and other expenditure since the opening.

The canal was a remarkable piece of engineering; the longest tub-boat canal in Britain, and that with the most inclined planes. It was a conception which outran the day-to-day reliability and strength of the materials then available for such mechanisms as the planes, far as they were from any industrial centre where repairs could be done or new parts made. The credit due to James Green must therefore be lessened by the sum of the practical operational difficulties that were encountered, and the maintenance costs of overcoming them.

On the Bude Haven section the sea-lock, after reconstruction in 1835, was 116 ft long and 29 ft 6 in wide, and would admit ships of 300 tons to the first ¼ mile of the canal, 10 ft deep, which formed the basin. Green's original plan, with which we have seen Kingdon experimenting, was that barges, 50 ft × 13 ft × 3 ft 6 in, able to carry up to 50 tons, should be taken out through the sea-lock and allowed to settle on the beach as the tide went down. They would then be filled with sand, brought in again on the rising tide, and towed to Marhamchurch. Probably the difficulties of handling the barges outside the sea-lock soon gave Lord Stanhope the idea of laying down an edge-railway* on the beach and up to the basin, where the barges, and also coastal craft, could be loaded. Thence the barges were horse-towed to Marhamchurch, except for a short time during which a steam tug was used. But this damaged the banks, and was withdrawn.

If the cargo were going farther, it was loaded straightaway into tub-boats at the basin. These were 20 ft × 5 ft 6 in × 20 in in size, carried some 4 tons each, and worked in gangs of six or eight. They were fitted with small wheels of 14 in diameter which projected from the sides, and ran in channel rails on the inclined planes, so that neither cradles nor caissons were needed. Such boats were only used on this canal and probably the Torrington, and were not altogether satisfactory:

> the boats whether in gangs of six or eight materially injure the slopes of the banks by their wheels, and when it blows at all with light boats it is impossible for the men to steer clear of the banks. The traders have readily come into a proposal to send at our joint expense men to check the boatmen, as they suffer equally with the works in their boats.[10]

There were six inclines on the canal, each with a double line of rails. The first from Bude was at Marhamchurch, 120 ft rise and 836 ft long; the second, 2 miles farther on, at Hobbacott Down (Thurlibeer), 225 ft rise and 935 ft long, or nearly 1 in 4. On the Thornbury branch there was a plane at Venn (Veala, Vealand, or Burmsdon) beyond the Tamar aqueduct on the east side of the river, which took the canal up another 58 ft in a length of 500 ft to its summit level, just before the Virworthy feeder led off to the north at Brendon Moor junction. The other three planes were on the Druxton branch, and lowered the canal from the height reached at the top of Hobbacott Down to the Druxton level. They were:

> Merrifield, 360 ft long, 60 ft fall;
> Tamerton, 360 ft long, 59 ft fall;
> Werrington (Bridgetown), 259 ft long, 51 ft fall.†

The boats were drawn singly up the planes by a chain passing round a drum at the top, but two or three empties were taken

---

* For information about this railway, see *The Railway Magazine*, 1917, p. 96, and 1922, p. 73. After the canal was closed it was still used to load coastal craft and railway trucks.

† It is interesting to note that in the Middle Ages Werrington belonged to Tavistock Abbey. Bude sand was regularly used upon its fields in 1350, and every year a train of six or seven packhorses went to and fro between Werrington and Widemouth in Bude Bay.[11] The canal therefore repeated this old traffic.

down together if required. The motive power in all cases was water, at Hobbacott Down the bucket-in-a-well system that was later to be tried on the Grand Western at Wellisford, and water-wheels elsewhere. Marhamchurch, for instance, was worked by an overshot wheel 50 ft in diameter, which raised a boat in five minutes start to stop, and used 32 tons of water to do it. At Hobbacott two buckets, each of 8 ft diameter, and using 15 tons of water, rose and fell in wells 225 ft deep. When the full descending bucket had reached the bottom of its well, it hit a striker, which operated a valve to allow the water to run out through an adit or tunnel to the lower pound. A 16 hp steam-engine was kept in reserve at this plane against accidents to the bucket machinery. At Hobbacott there was also a small 9-ft water-wheel, which worked a turning lathe in the canal blacksmith's shop.

When the planes worked well they were efficient. For instance, the resident engineer reported on 2 July 1827 of the Hobbacott Down plane that 'The bucket machinery is in perfect repair and works extremely well. 59 boats were taken over this plane on the 25th. ult. in 7 hours.' On the other hand, breakdowns were frequent, and each of them usually led to a stoppage. Apart from replacements of timber-work and rails, the following accidents took place in the eight months November 1825 to June 1826:

November 1825	Venn main chain broke.
December 1825	Boat broke loose on Werrington plane.
do	Gear wheel broke at Werrington.
do	Bucket chain broke at Hobbacott Down; bucket damaged.
January 1826	Bucket chain broke at Hobbacott Down; bucket damaged.
March 1826	Boat broke loose at Hobbacott Down.
April 1826	Bucket chain broke at Hobbacott Down; bucket broken.
May 1826	Both bucket chains broke at Hobbacott Down; bucket damaged.
do	Main chain broke at Hobbacott Down, and boat ran away.
do	Main chain at Venn broke.
do	Main chain broke again at Hobbacott Down.
do	Chain wheel shaft broke at Tamerton.
June 1826	Gear wheel broke at Marhamchurch.
do	Main chain broke at Marhamchurch.
do	Main chain broke twice at Hobbacott Down.

The men in charge of the planes, and especially at Hobbacott Down, must have earned the 2s to 2s 6d a day that they were paid.

The same sort of accident was still happening ten years later, for when W. A. Provis visited Hobbacott Down plane in 1836, he wrote:

> I found that the inclined plane was being worked by a Steam Engine. Some weeks previous to my visit (which was on the 16th. May) the Endless Chain had given way in consequence of which the Boats had run down the Inclined plane, and the loaded bucket had fallen to the bottom of the well and been smashed. . . . I have been informed that several accidents had occurred before the one which I have described.[12]

James Walker, however, wrote in 1838:

> Every means appears to be taken to avoid accident; the machinery though complex—and perhaps necessarily so—is ingenious. It has, I understand, given trouble but not lately, and it does not appear with care which is indispensable, likely to do so to any extent, although with such heavy machinery accidents are unavoidable occasionally.[13]

But Hobbacott was much more costly than the other planes. For the year ending 31 March 1839, for instance, out of total canal maintenance expenses of £1,221, Hobbacott took £212.

The life of the resident engineer on such a canal was a strenuous one. On 4 January 1827 the chairman said of him:

> The last month has been a period of misfortunes, and they have followed each other so quickly that Mr. Honey and his assistant have had most severe duty to perform. I beg you gentlemen to judge the exertions of the former when I state that he rode three hundred miles in five days, and I am astonished at his having been able to stand up against such fatigue.[14]

Information about the affairs of the company is fragmentary. In its first few years accidents were so frequent that the income did not cover the expenses. For instance, on 14 January 1827, the chairman wrote: 'Whilst these repeated casualties occur I despair of any benefit resulting from the Canal to the Proprietors, even if it can pay its way,' and in May he reported:

> It is with the deepest regret that I am obliged to call your attention to the Contingent account No 1, amounting to £2,747 18s 5d, a sum far exceeding the amount of tolls for the year. . . . The machinery and canal banks have failed repeatedly and the consequent stoppage of trade has materially affected the revenue. Under the first Head the chains have parted, buckets have been broke, metals on the several inclines have given way. This occasions much damage to the boats passing the machinery by irregular movements . . . the canal banks have been found

very leaky. . . . In many instances the banks have settled and the puddle being below the level of the water when the canal is full it escapes through the banks occasioning much damage to them.

In the autumn of 1830 John Honey was wringing his hands over delay in the delivery of new iron wheels for the bucket machinery at Hobbacott Down, and the consequent coal consumption by the steam-engine. He was at his wits' ends to pay taxes and his men, and because tolls were too high, 'it appears to me that the Farmers who are the only support of this canal, have . . . totally set their face against it'.[15] At this time the company was employing 19 men, a lock-keeper, wharfinger and tramway man at Bude, mechanics at five out of the six planes (Venn seems not to have had one, and was perhaps worked by the boatmen), two masons, a smith and seven labourers.

In the following spring he wrote:

It being now Market Day at Stratton where the Mechanics and Labourers get their supply for the ensuing week, I have had almost the whole of them in the Co's employ last night & this morning by day break request to be paid their Wages, and otherwise they inform me they cannot procure necessities for their support. And I assure you, Sir, it does in a great degree lessen the influence which could be operated on them, on Account of their Wages being kept unpaid for so long after it becomes due. . . . For my own part I know not what to do. I have been call'd on this morning by no less than 10 persons for taxes. I've not a farthing to pay them, and worse than all this, there is no Trade whatever on the Canal, and only 177 Boats have been taken over the Planes from the 1st. of this month to the present time. This trade will not pay for the coals and labour in working the Engin and other Labour in passing the different Planes.[16]

And then:

I've done all I can with the traders and farmers, and the former inform me that they cannot sell the sand if they take it up, if they are to pay the tolls now demanded. The latter informs me that they cannot purchase the sand at the price the traders ask for. Therefore it has been the means of driving many farmers to the beach[17]

to get their own sand. The iron wheels arrived, but then the boiler was found to be rusted out, so that Mr Honey hoped nothing would go wrong with the buckets.

In 1831 tolls were lowered, and trade began to pick up. But

John Honey, who had added engineering to his former clerkship, was replaced by Joseph Cox, one of James Green's men, in March 1832. He was an earnest man who had earned praise in the past, but he so consistently looked on the dark side of affairs, and grumbled so perpetually, that he probably got on the nerves of the chairman and committee.

Afterwards things got better. For the year ending 31 March 1838 the tonnage carried was 59,620½, 54,016 being sand, and the revenue from tolls, harbour and basin dues and rents was £4,339, and the expenses £2,339 before charging interest. No dividend had yet been paid, and the company owed the Exchequer Bill Loan Commissioners £21,037 for principal and interest, and £3,200 of other debts. It was said at this time that sand at Launceston was only one-quarter of the price it had been before the canal was built, and that the value of farms near it had improved, together with their rentals. The tonnages actually carried, however, were far below expectations. In 1818 it had been calculated that 28,038 tons of sand were already being conveyed into the interior by land carriage, and a total of 312,500 tons of sand and 2,530 tons of other goods was not thought too optimistic an estimate of carryings on the canal. In fact, in 1848 the total tonnage carried was 52,501, which yielded a gross revenue of £3,533, and this tonnage figure remained fairly consistent throughout the canal's history. Most of it was sand off the beach, but coal, culm, timber, salt and miscellaneous cargoes were brought to Bude by small coasting craft, which carried grain and bark away, or left in ballast. In 1832, for instance, 270 craft entered (against 129 for 1825) averaging 40¾ tons. The largest was 125 tons.

Three events need to be recorded. In 1835 the engineer James Meadows Rendel enlarged the sea-lock and the basin. In 1836 an Act was passed to authorize the building of a railway from Tremoutha Haven near St Gennys, midway between Bude and Boscastle, to Launceston, and to build a harbour at Tremoutha, as part of a new town to be called Victoria. No action followed: if it had, there would have been serious competition with the canal for goods except sand.[18] As it was, a railway did not reach Launceston till 1865, or Holsworthy till 1879, and competition therefore came late. Lastly, in February 1838 the breakwater at Bude was badly damaged by a great storm. The company thereupon asked the Exchequer Bill Loan Commissioners to agree that their trading surplus could be used to rebuild it instead of to pay interest on their loan. The Commissioners agreed, but for a time it was found impossible to raise the sum needed. Finally it was decided to build a smaller and cheaper breakwater than the old one, and this was finished at the end of 1839 at a cost of £2,000. It was again reconstructed in 1856.

In 1839 the debt to the Commissioners, including unpaid interest, reached its highest point at £22,427; by 1864 it had been reduced to £3,000, and in 1870 it was paid off. Takings never exceeded the £4,716 of 1841, long before the railways came to compete. The basic trouble about the canal was a maintenance cost far above what had been estimated; therefore tolls had to be kept at a level where the profits to traders were so small, if prices to farmers were to be kept attractive, that the business was hardly worth undertaking. Any effort to raise prices meant that the farmer fetched the sand himself. Because the company did not themselves act as carriers, they found it necessary to lend capital to others to keep their trade, while at the same time doing all they could to keep costs down. Their largest profit was the £1,493 of 1859. But takings were falling off during the fifties, due to the improvement of roads and the introduction of artificial fertilizers. The average of £4,142 for the three years 1840–2 had become £2,982 for the four years 1861–4. For this latter period harbour dues averaged an additional £220, and average tonnages were:

	Tons
Sand	39,616
Coal	6,017½
Culm	2,517
Goods	2,585
	50,735½

It was not until 1865 that the broad-gauge Launceston & South Devon Railway reached Launceston from Tavistock, or 1874 that the South Western was connected with this line at Lydford. In 1879 a line from the South Western reached Holsworthy by a branch from Meldon near Okehampton, and was extended to Bude in 1898.

In the year that the Launceston & South Devon line reached Launceston, an Act was passed for a Bude Canal & Launceston Junction Railway Company, with a capital of £20,000, to join the canal to the railway. The engineer was J. F. W. Featherstonhauge, whose plan proposed a line 2½ miles long from Druxton Canal basin beside the Tamar to join the Launceston & South Devon near Lower Bamham with two junctions, one towards Launceston, the other towards Tavistock. It was a time when there was talk also of bringing the standard gauge to Launceston and on to Wadebridge, and provision was made for a possible future link with it. The Act, which had been promoted by Launceston interests, and not directly by the canal company, was not carried out. Instead, tolls had to be reduced on the Druxton line, with the result that in 1863, 37,742 tons of sand over the whole canal paid £2,151, and in 1865 54,770 tons paid £1,684. Some of the coal trade was lost, and most of that in culm.

In the four years ending with 1876, the average tonnage on the canal had increased to:

	Tons
Sand	48,829
Coal	4,223½
Culm	782
Goods	1,311
	55,145½

Canal tolls had averaged £1,897, and with other sources of revenue such as harbour dues, which yielded £152 pa, there was a small surplus which, now that the debt had been repaid, could be distributed. The company's first dividend, of 10s per £50 share, was paid for 1876, fifty-three years after the canal had opened. There were to be only seven more: 3s 4d for 1877, 5s each for 1878 and 1879, 6s 8d for 1880, 3s 4d for 1882, 2s for 1887 and 2s 6d for 1899.

James Sleeman worked as a handyman on the canal in the seventies and eighties. His diary[19] records his work, much of it on the planes. Here are a few entries:

3 Jan 1879	Marhamchurch plane altering gear
28–31 March	Sawing brake for Tamerton
1 April	At Marhamchurch Plane about Gear
10 June	Sawing for Werrington paddle frame
25–30 June	Hobbacott Down fixing framing to governors
3–4 July	Hobbacott Down about framing. Repairing buckett
7–8 July	Sawing and preparing chain wheel ring for Marhamchurch
8–11 August	Sawing for Tamerton Frame around shaft

In 1877 one of the two principal traders on the canal closed down, and the company bought his stock of 56 boats, 7 barges and 8 waggons, later hiring them to Vivian & Sons, their only remaining customer. In that year they reduced the toll on culm on the

XIII. (*above*) Tavistock Canal branch to Millhill in 1967. This passed under the bridge. The likely site of the inclined plane is between the bridge and the first building on the left, the upper canal then running on the discernible line behind the cottages to the quarry, beyond the picture to the right; (*below*) the Weare Giffard inclined plane on the Torrington Canal in 1954. The embankment of the later railway, in the background, intersected the upper part of the plane

XIV. Bude Canal: (*above*) The basin about 1890, showing tub-boats being loaded; (*below*) a portion of the canal, now maintained as a water-channel; near Venn (Veala) inclined plane

Druxton line to ¼d per ton per mile in a last effort to keep this traffic, and decided to sell the worn-out engine and boiler of Hobbacott Down plane, now to have no alternative to its temperamental buckets.

In 1878 the committee wondered whether to build a light tramway from Druxton to Launceston, and in 1879, a few months after the railway to Holsworthy had opened, reduced their coal tolls to Stanbury Cross and Blagdonmoor. Oddly enough, a trade developed in sand brought by canal to Stanbury Cross and then transferred to rail, but it did not last long. The future was not to prove hopeful. From 1880 onwards sand-carrying slowly decreased: it was 37,844½ tons in 1881, 24,893 in 1884, 23,687 in 1887, and 16,403 in 1890. Coal fell by more than a third after the opening of the Holsworthy railway, from 5,561½ tons in 1878 to 3,232½ tons in 1880, and then held steady to 1890. Culm disappeared after 1878, but miscellaneous goods actually increased, from 1,934¾ tons in 1878 to 2,764 in 1890. The total traffic fell in twelve years by more than half, from 48,568¼ tons in 1878 to 22,315 in 1890. Revenue also slid down, averaging £1,256 in canal tolls and £136 in harbour dues for the four years ending 1884, and then falling further, so that in only three of the nine years ending in 1891 did it exceed expenditure.

In October 1882 the committee wrote to Vivian's that trade was decreasing each month, asked for their suggestions, and added in December that they had learned of a rise in Vivian's freight charges for sand, and hoped it would be reduced. Soon afterwards they were disturbed to hear of a proposed extension of the Holsworthy railway to Bude. They saw no chance of successful opposition, but concentrated on getting protective clauses. On top of this Vivian's wrote in February 1884 that they proposed to give up trading at the end of the year. Reckoning that sand would never again be profitably carried in big enough quantities to compete with rail-borne lime and artificial manures, the shareholders then decided to abandon the canal and wind up their company.

At which point Vivian's had second thoughts, and asked what traffic was required to cover the canal's expenses. They were told that to keep it open for 1885 the company would want £1,000 in tolls, plus boat hire and harbour and basin dues. Because they did not clearly accept this offer, the abandonment bill went ahead, only to be complicated by an offer by Edwin Chamier of Stratton to arrange the purchase of the company for £5,000. At first this was refused. Then, some shareholders showing themselves willing to sell, the bill was withdrawn at a cost of some £600 to the im-

L

pecunious company, and Chamier's offer was accepted, only to
have it withdrawn.

In 1885, however, Vivian & Sons agreed to guarantee £1,100
pa of revenue, and the committee decided to keep the canal going
for the time being. They agreed to pleasure boats being put on the
Bude section at a yearly toll of a guinea each, and raised tolls on
miscellaneous goods from 1d to 1½d. They also turned their minds
to the possibility of selling water from their reservoir at Alfardisworthy, at first to Plymouth, and later to Launceston and Bude.
By 1888 they had again decided to abandon the canal, and agreed
with Vivian's that the latter's trading should be on a temporary
basis until the end of 1890. But serious talks about supplying
water to Stratton and Bude went on until early in 1891 the company dropped their old bill and substituted another.

The Bude Harbour and Canal (Further Powers) Act[20] of 1891
empowered them to close the canal from Marhamchurch plane to
Brendon Moor junction, together with the Druxton and Holsworthy branches. The sections from Bude to the bottom of Marhamchurch incline and from Brendon Moor junction to Tamar
Lake were to be retained as navigable waterways. It was intended
that water from the reservoir would be brought through the Virworthy branch and the main canal to the top of Hobbacott Down
plane, and thence supplied to Stratton, Bude and other nearby
places, for which purposes they were empowered to raise £10,000.

All the men working on the Druxton branch, except two on the
planes, had been dismissed in April, and now the company gave
notice that this and the Holsworthy line would be closed on or
before 14 November. The committee began to dispose of the land
of the old line, held sales of equipment at Druxton on 14 September and Hele bridge on 26 October, and began to break up boats.
At the same time they reduced tolls on sand and miscellaneous
goods on the Bude–Marhamchurch section.

The Act had been obtained upon an understanding reached between Mr Edward Mucklow, a leading committeeman and shareholder, and the Board of Guardians for the Stratton Union, the
local sanitary authority, which had later been confirmed by an
agreement of 7 April 1891, that the latter would take the company's water to supply their area, with a minimum payment of
£50 pa, leaving the company free to supply elsewhere. But nothing
came of it, seemingly because the company could not raise the
funds to deliver the water in proper condition at Hobbacott
Down. In February 1894 the committee minuted that they understood 'the Bude and Stratton Sanitary Authority have now defi-

nitely given up any intention of obtaining a water supply for their District from the Canal or Reservoir'. The committee recommended sale as a going concern or abandonment of the remaining works.

In 1895 the company tried unsuccessfully to sell itself to the London & South Western Railway, whose extension from Holsworthy to Bude was now imminent, and then in 1896 opened negotiations with the newly-formed Stratton Rural District Council for the sale of their reservoir and water rights, though they now refused to consider supplying water themselves. Edward Mucklow then offered to buy its assets for £3,500, intending to form a syndicate to supply water to the district, at the same time maintaining the breakwater, harbour and basin at Bude and the lower reach of the canal, the syndicate to pay the costs of an Act. The committee were willing to agree if the price were increased to £3,622, or £2 per share, but the proposed syndicate could not be formed. To add to the variety, in March 1897 Messrs Dickson & Co. wrote offering to buy the whole undertaking for £5,500, but there is no indication of what they proposed to do with it, or who they were.

From all sources revenue in 1897 was £364, and in 1898, £359, the former year showing a profit of £100, the second of £62. Canal tolls were £36 and £9 respectively. It seems unlikely that there was any canal traffic after 1898, while craft entering the harbour dropped sharply as soon as the railway opened to Bude in August 1898, though a small transhipment trade from ships to rail developed. Then, in February 1899, the company reached agreement in principle with the Stratton and Bude Urban District Council* to sell their undertaking for £8,000 which, after some compensation had been paid, gave each shareholder a return of £4 per £50 share. The purchase was authorized by the Stratton and Bude Improvement Act of 1901,[21] and the company formally handed over on 1 January 1902.

Instead of water being run to the top of Hobbacott Down, as had been intended in 1891, the Council piped it from Venn (Vealand), using only the canal section of 4¾ miles thence to the reservoir as an open channel. The rest was not used, and was abandoned as a waterway by an Act of 1912.

By the Bude-Stratton Urban District Council's Act of 1960,[22] the council were authorized to dispose of the land between Venn (Vealand) and Marhamchurch planes (except the Tamar aqueduct and Burmsdon bridge), to close the locks of the broad canal

* It was renamed Bude-Stratton Urban District Council in 1934.

(which had in fact been done without authority in 1924), and make Falcon bridge a fixed instead of a swing bridge. In 1967 they announced their intention to seek a bill to substitute a weir and spillway for the gates of the entrance lock.

The reservoir, now called Tamar Lake and a protected area for birds, and about 5 miles of the old waterway to the top of Venn (Vealand) plane are still used for water supply. The reservoir and the canal as far down as Burmsdon were transferred to the North Devon Water Board in 1967. Thence to Hele Bridge (Marhamchurch) several lengths have been sold. Below Hele the Canal is still owned by the Bude Stratton Urban District Council. The edge-rail tramway at Bude carried sand from the beach to the basin until August 1942, when it was superseded by the motor lorries that still supply local farmers. Pleasure boats are still rowed along a stretch of the old broad waterway, the last traffic of the Bude Canal.

CHAPTER XI

Other Canals of Cornwall

++++++++++++++++++++++++++++++++++++◆++++++++++++++++++++++++++++++++++++

Parnall's Canal

A CANAL about ½ mile long seems to have been made by Mr Parnall, perhaps about 1720, as part of the Carclaze mine near St Austell. The boats were filled with 'tin-stuff' in a tunnel of the mine, and then moved to the stamps near the bottom of the hill in which the mine was cut. Here the boats were unloaded into carts, perhaps by up-ending them with the help of a windlass and allowing the contents to run down a slide.

The boats were flat-bottomed, 6 ft long, 4 ft 6 in broad, and 1 ft deep, and were worked in trains. Parnall's Canal was closed by the falling in of the tunnel about 1732, so imprisoning the eighteen or so boats used. They were rediscovered about 1850.[1]

St Columb Canal

If there were such a slide, Mr Parnall's canal may have given John Edyvean the idea of the St Columb Canal. This was authorized in 1773[2] from Mawgan Porth through the parishes of Mawgan, St Columb Major and Little Colan to the sea again at Lower St Columb Porth, the main purpose being the import of sea-sand for manure, for which it 'will afford the means of improving many thousand acres of barren and unprofitable Ground',[3] and of coal, and the export of stone.

The canal was projected by John Edyvean, who proposed to build it at his own expense. According to the engineer, John Harris, who gave evidence before the Lords committee,[4] its length was to be about 30 miles (the committee clerk must have misheard thirteen as thirty), at an estimated cost of £5,000 to £6,000, which would have been low for a tub-boat canal of even 13 miles. But seemingly Edyvean had a bigger waterway in mind, for the Act authorized him to charge boats on a minimum of 20 tons of cargo, which looks as if at least 25-ton craft were envisaged. Tolls were to be a maximum of 1s a ton for sand and 2s 6d for merchandise;

Harris reckoned that sea-sand could be carried at half the cost of land-carriage.

Two portions, seemingly on tub-boat scale, were built. One ran from the sea at Trenance Point above Mawgan Porth at a height of about 200 ft for some 4½ miles past Porth Farm, Moreland, Lower Lanherne, New Farm, Trevedras and Bolingey to end about ½ mile below Whitewater, where there was an intake from the river. From the end of the canal at Trenance Point an inclined plane was intended to raise sand from the beach. The second ran from Lusty Glaze, south of St Columb Porth, at about 100 ft past St Columb Minor to Rialton and as far as the road junction leading to Rialton Barton. At Lusty Glaze a cut for an inclined plane was made down through the rock to the sea.[5]

This second section was probably started first. An advertisement of July 1773 described it as 'now cutting', and another of August for Colan Barton says: 'The premises lie very convenient for sea sand . . . next year the intended navigable cut or canal will be perfected, and run through part of the barton.' Another of August 1776 for Nanswhyden barton near Colan says: 'next summer a navigable cut or canal will run through . . . the premises', and a year later one for Bejowan refers to it being 'very conveniently situated near Mr Edyvean's canal for bringing sand and other manure thereon'. This could be taken as implying that the canal was open; if so, this was not the case in January 1779, when the same property was again advertised without the canal reference. I think it is very doubtful, however, whether this section was ever finished; one hazarded reason for failure was that Edyvean could not get some sandy soil to hold water.

He started the Mawgan Porth line soon after beginning that at Lusty Glaze. An advertisement of June 1775 concerning a farm that had 3 tons of sand a day brought by ox or horse waggons says: 'A navigable canal is now making nigh the premises, where sand may be had much easier.' A year later another appeared: 'Wanted, on the Maugan Canal, Cornwall, fifty able men, that can work with shovel, or pick, at Fourteen pence per day. Better encouragement on tutt work.'*[6] It is thought that this section of canal was opened and used for two or three years.

The following description of an inclined plane, written some twenty-five years later, is attributed to that at Lusty Glaze; if there were two, it is likely that they resembled each other. Alternatively, it may really describe that at Trenance Point, which is attested by one who saw it:[7]

* Payment by the piece or measurement.

The canal was a narrow one, and at its west end, the cliff was with great labour hewn down, into a steep inclined plane, that was covered with planks; the canal was conducted to the very top of this plane, and the boats which were rectangular ones, were brought, when loaded with stone, to the termination of the canal, where they were fastened by a sort of hinges; strong ropes were then attached to the stern of the boat, and by means of a wheel and drum, worked by a horse-gin or wem, the boat was hauled up on end, and the stones were thereby shot out, and rolled down the plane to the strand below, from whence boats conveyed them on board the ships. The same wheel and drum was adopted for drawing boxes of coals or shelly sand up the plane, in order to their being loaded into the returning boats.[8]

Little is known about Edyvean. He is described as of St Austell. Given the assumption that John and Joseph Edyvean, later associated with the Polbrock Canal, were relations, then it is possible that our John Edyvean was a farmer, for in 1795 the barton of Boconnion, then occupied by Joseph, was advertised to be let, and described as convenient to Bodmin, St Austell and St Columb, and within a short distance of Polbrock, 'whence sea sand may be had at a small expense'.[9] If so, the elder John's interest in sea-sand is clear. He died in the 1780s, and a kindly man wrote pseudonymously about him in September 1791:

John Edyvean . . . was born to affluent circumstances, but dissipated his wealth in pursuits that had for their object the good of mankind, although he failed to get their sanction and support. . . . About the year 1777, he laid before a county meeting of Cornwall, the plan of a canal for traversing the whole kingdom without a single lock, by means of inclined planes, but it was rejected as wild and chimerical.

Later in life, the writer goes on, he planned the St Columb Canal. In the year 1779, he had finished the canal up to the town of St Columb. . . . In that year, I went with some friends to visit this work. We overtook the poor old man, groping his way by the side of his canal, and leading a miserable little horse in his hand. We joined him, and he conducted us to all the parts of this ingenious work, with the intelligence of one who had formed the whole, inch by inch, and this alone can account for the ease and safety with which, in his blind state, he passed through every part of it. We dined together, and he gave us a little history of his life.[10]

With the help of the large-scale plans, it is still possible to trace much of the course of both canals.

Liskeard & Looe Union Canal

The building of the St Columb Canal, and the Act for that at Bude, in the summer of 1777 suggested to a gentleman of Liskeard the possibility of such a manure canal to carry sand and lime from Looe, a trade then carried on by pack-horses. He consulted Edmund Leach, who proposed a small canal from Bank Mill bridge, 2½ miles from Liskeard, to Sandplace 2 miles above East Looe, 15 miles long against 8 miles by the river, and with two planes. The cost was estimated at £17,495, and the receipts at £2,500 per annum. Leach suggested also that if successful the line could be extended to St Germans to open a trade with Plymouth.

Nothing was done, and it was to be twenty years before the scheme was revived about 1800, when surveys were made of lines from Liskeard to Looe and from Liskeard to St Germans, but the cost and difficulties were thought too great.

In 1823, after 'a numerous and respectable meeting' had been held at East Looe in August, with the mayor in the chair, and approved making a canal or railroad,[11] a committee was formed 'for ascertaining the best Mode of making an improved Communication between the Towns of East and West Looe and Liskeard'.[12] James Green was called in and asked to report upon the possibilities of a canal, a railway, and a turnpike road. He rushed through the work, relying partly upon levels taken earlier, presumably in 1800, and finished it in a fortnight. He gave his opinion that all three possibilities were practicable, but in the case of a canal, added: 'The descent of the valley is too rapid to admit of a Canal being formed on the common principle of Lockage.' He went on to say that the expense would be too great, and the time taken to pass the locks too long, and the water supply insufficient, and ended: 'I am therefore of the opinion that the major Part of the Ascent must be overcome by means of Inclined Planes, and that the Canal should be made for Boats of Four Tons Burthen, which are well calculated for the Trade likely to pass over it.'

The lower part of the canal from the entrance at Terras Pill on the Looe river past Sandplace to Causeland was to be larger in size, 28 ft wide at top and 4 ft deep, and locked, in order that the limestone barges could reach Sandplace,* where limekilns were to be

---

* Sandplace was the spot to which sand was taken by barges. The barge crews dredged the sand from under-sea banks by towing canvas bags fitted over iron hoops behind their craft. Until the canal was built pack-horses loaded it there into panniers with flaps at the bottom so that the sand could be emptied directly on to the fields.

built, at all tides along an easier course than the winding river. He did not think a suggested extension to Steps to be justified. Above Causeland, there was to be a small canal, 19 ft top width and 3 ft deep, along which the 4-ton craft would be worked in trains of up to ten, and two inclined planes.

The resemblance to his Bude Canal scheme is obvious. He estimated the cost at £14,000. The traffic envisaged by Green was limestone and culm to burn at the kilns at Moorswater, coal, iron, timber and merchandise for Liskeard, and agricultural produce downwards to the sea. In other words, the canal was thought of primarily as an agricultural and coal-carrying waterway, though when it opened tolls were fixed for copper and other ores, and clearly some traffic was expected from the St Cleer, Linkinhorne, and St Neot mines.

Meetings early in September at East Looe and Liskeard considered the report, and decided to seek a bill, Green being asked to do the plan, which was deposited in October. Green then departed, there were second thoughts, and in the end the local men proceeded to ignore his advice. Maybe they knew the heavily locked canals of South Wales, and preferred their known difficulties to the unknown troubles of inclined planes and tub-boats. The line was re-surveyed by John Edgcumbe,* Robert Coad and Thomas Esterbrook and estimated at £12,578; then in 1825 an Act was obtained for a locked canal, 26 ft wide at surface, and 4 ft deep from the eastern side of the River Looe at Moorswater, 1½ miles from Liskeard, to cross the river near Trussel bridge and thence run on the west side to Corgorlon bridge, whence a wider and deeper canal would continue past Sandplace to Terras Pill. There were to be 24 smaller locks and a larger river lock in 5⅞ miles, with a total rise of 156 ft. The authorized capital was £13,000, in £25 shares, with power to raise another £10,000 if necessary. Of the sum needed, £10,500 had been raised by that time from 134 shareholders, the only big subscribers being Rowland Stephenson (£1,500), Messrs Glubb & Lyne (£500), Hon. John Walpole (£625), Sir Manassah Lopez, Bt. (£525) and John Buller (£500). There was opposition, for the first shareholders' meeting after the passing of the Act thanked the Clerk, Peter Glubb, 'for the zeal at all times manifested by him in support of the project for this Canal to which in a great measure may be attributed the success

---

* Edgcumbe was a Liskeard engineer who, many years earlier, had ended an advertisement for threshing and winnowing machines, flour mills and cider presses by 'Canals cut on a new patent plan eighteen feet wide, and three deep, with all apparatus complete, in good ground, £900 per mile. Likewise Surveys the Lines of Canals'. He finished by offering Edgcumbe's New Patent Inclined Plane.[13]

of the arduous struggle', and expressed 'their admiration of the Talent and Spirit displayed by him in conducting their case before a Committee of the House of Commons where he triumphed over a powerfully combined opposition without calling in the aid of Counsel'.

Robert Retallick was also thanked for 'keeping alive the project when some of its early friends deserted it'.[14]

The engineer was now Robert Coad, with Robert Retallick as Superintendent of Works. They proceeded to lay out the line, start cutting, and build the first lock, the foundation stone of which was laid by the Rev. Mr Jope. Meanwhile we may note an early health insurance scheme, for it was decided 'That Mr. Robert Rean of East Looe be Surgeon and Apothecary to attend the Laborers working on this Canal and that six pence per month be deducted from the wages of each person employed, as a Remuneration to him for his Services.'[15]

Early in the following year there was some nervousness at the qualifications of the men who were overriding Green's recommendations and a motion was put to the shareholders' meeting: 'That a civil engineer properly qualified be called in to view and examine the lock now reported to be completed and to report thereon and also to examine the works already proceeded in and also report if the Superintendent of the Works Mr. Retallick and Mr. Robt. Coad are sufficiently qualified to proceed without further assistance.' The motion was lost by 49 votes to 10. Work then proceeded, the committee busying itself to get calls paid and collect arrears. The canal was partially opened on 27 August 1827 and fully during March 1828, and in the following year the company built a new road from Moorswater to Liskeard to avoid the steep hill on the old road. This canal was remarkable for paying a dividend of 6 per cent for 1829, its first full year of working, and for a construction cost not greatly in excess of the estimate; £17,200 against £13,000, but including the new road and £600 compensation paid to a landowner under a special agreement. After 1829 the dividend fell back to 5 per cent, which was maintained in most years, though often paid late because of the difficulty of getting money in. Most of the traders had monthly accounts, but the kiln owners only paid once a year, at Christmas, for their limestone and culm. There is a hint of trouble, however, in a letter of 1829 from John Buller to his lawyer, which refers to the canal committee as 'not composed of Men of Education, but I believe of a very artful and cunning description'.[16] But we do not learn what the trouble was.

So far the canal had been a reasonably prosperous little affair, but now a new factor arose. At the beginning of 1836 the Committee reported to the shareholders that because of unprecedented agricultural distress the carriage of lime had considerably decreased, but that

they are enabled to congratulate the company on the prospects of an increase in tolls for the ensuing year from the circumstance of the neighbouring mines being about to be prosecuted with spirit and capital sufficient to ensure their success in which case there is every probability that the produce of the said mines will be shipped at Looe thereby affording back carriage to the boats employed on the canal which have now scarcely any back carriage.[17]

In 1837 the South Caradon copper mine was opened, and in 1840 that at West Caradon, the ores being carried by wagon to Moorswater and then shipped on the canal. To help this trade the tolls on copper were reduced from 1s 6d to 1s 3d per ton in 1838, and again to 1s per ton in 1842. Meanwhile, however, other trade was falling off, the traders saying that the canal charges were too high compared with those for land carriage for St Germans, and the company that the traders were being disloyal, as

It behoves therefore all persons interested in the Trade on the Line of the Canal to do their best to support it, for it is only by general support that it can be protected and preserved. Let the canal be destroyed and the consumers of lime and coals would soon have to pay the old prices from St Germans.[18]

In 1843, however, the toll on coal was reduced from 3s to 2s after further complaints that the coal dealers at Looe and St Germans were underselling those at Moorswater.

The output from the Caradon mines was now beyond the capacity of road wagons, and in 1843 the Liskeard and Caradon Railway was authorized, some of its directors being also directors of the canal company. A line from the two mines to Moorswater was opened in 1844, and a branch to the granite quarries at Cheesewring in 1846. These were worked by horses till 1862. Before this, however, the railway had been in financial trouble because some of its shares had been forfeited to avoid payment of calls. The canal company had thereupon taken up 25 shares at a cost of £625, and the mineowners the rest.

The canal company was still doubtful about its future; in December 1845 it resolved to oppose the Cornwall and the Devon & Cornwall Central Railways (only the former was afterwards authorized), and at the same time

Resolved that notwithstanding the above resolution it is advisable to ascertain if either of the proposed county railway companies will treat with the said canal and Liskeard and Caradon Railway companies for their respective properties . . . and if so on what terms.[19]

In 1846 an agreement* was in fact made with the Cornwall Railway, for that Company's Act of the same year gave the railway power to lease or buy the canal and the Liskeard & Caradon Railway. The Cornwall Railway, which included the Royal Albert Bridge over the Tamar, took many years to build, and was not finished to Truro till 1859. By then much had happened, and the option was not used. Meanwhile, on 5 February 1845, a public meeting at Liskeard had decided that the port of Looe was inadequate for the traffic offering, and that it should be improved in connection with the canal. From it followed an Act of 1848 establishing Harbour Commissioners, one of whom was to be the canal company's treasurer.

From 1846 the output of the mines steadily increased the traffic on the canal, enabling the mortgage debt to be paid off in ten years without affecting the regular 5 per cent dividends, and leaving an increasing balance in hand. In 1849 the tonnage was 24,000 tons, and in 1854, when it reached 36,000 (16,000 of which, mostly minerals, being downwards),† there was a proposal to extend the canal nearer to Looe to cope with the 'daily increasing traffic'. The upward carriage was also benefiting by the carriage of coal to the mines, and by 1856 the canal was working to capacity. In that year it carried 48,000 tons, mostly of coal, copper ore, limestone, and granite; it had receipts of £2,560, and made a net profit of £1,783. In October of the following year the directors sent a circular letter to the shareholders, in which they said traffic had reached the capacity of the canal owing to the impossibility of increasing the water supply, that even with the existing traffic water shortage made the canal difficult to work, and that goods from the Liskeard & Caradon Railway were being diverted to Calstock because of the congestion. They went on to say that the Cheesewring Granite Company were proposing to build a tramroad to the L. &C.R. still farther to increase their output, and that it cost from fourpence to sixpence per ton to transfer goods at Moorswater between the railway and the canal. They therefore proposed the building of a

---

* The Cornwall Railway had proposed to build a branch to Looe. The canal company opposed this, and obtained the insertion of the clause referred to.

† In the same year of 1854 the Liskeard & Caradon Railway carried 9,815 tons of copper ore and 3,364 tons of granite, practically all of which probably passed down the canal.

railway from Moorswater to Looe, but did 'not suggest the destruction or abandonment of the Canal, but that the one should be auxiliary to the other'. The line was to run alongside the canal to Terras Pill and then follow the eastern shore of the estuary to Looe. It was proposed that the canal shareholders should finance the railway building by utilizing the full borrowing powers of the original canal Act, and by using their now considerable reserves. The estimated cost was £11,000 exclusive of rolling stock and parliamentary expenses, the figure being so low because most of the necessary land was already owned by the canal company.

The shareholders decided to go ahead at once. The Looe Harbour Commissioners agreed to provide a terminus and quays for the line, the Granite Company decided to build their branch,* for which the canal carried the rails free of toll, the Act was obtained in 1858 and construction began, granite blocks from Cheesewring being used as sleepers. The engineer was Silvanus W. Jenkin. The cost, however, proved greater than the estimates and had reached £20,140 by the time the line was opened for goods traffic on 27 December 1860. The expense having been so great, the getting of rolling stock was a problem, and in May 1860 an agreement was made with the L. & C.R. by which the Liskeard & Looe acquired an engine and the other railway the trucks.

The Liskeard & Looe Railway had a greater capacity than the canal, but its cost of construction and working expenses, including those of operating which on the canal had been borne by the carriers, meant that the profits were little higher than before. The following comparison is interesting. It should be remembered that the receipts for 1863 include carrying receipts as well as tolls, those for 1858 only tolls.

	1858 (canal)		1863 (railway)	
	Tons	£	Tons	£
Coal	12,966	934	18,854	2,222
Copper	17,238	862	27,252	3,193
Granite	5,785	217	7,168	717
Limestone	5,214	124	4,159	208
Others	3,301	153	4,779	456
	44,504	2,290	62,212	6,796

The increase in tonnage of about 30 per cent only gave the company an increase in profit from £1,822 in 1858 to £2,124 in 1863. Indeed, while the old shares continued to get their 5 per cent

* The Kilmar Railway.

dividends, the new shares got nothing till the loans were paid off, and never reached the same level. Here are figures for the last seven years of full canal operation:

Date	Tolls £	Profit £	Tonnage Tons
1854	2,024	1,447	—
1855	2,311	1,658	—
1856	2,461	1,783	—
1857	2,419	1,852	—
1858	2,290	1,822	44,505
1859	2,525	1,978	48,193
1860	2,303	1,639	45,555

In 1862 the Liskeard & Caradon Company agreed to work the line, and the one engine was sold. The peak of copper production had, however, been reached. By 1886, seven years after the line had begun to carry passengers, the Caradon Railway was in the hands of a receiver, and its mounting debts to the Liskeard & Looe prevented that company from itself being more prosperous. In 1901 a connection was made between the Liskeard & Looe (it was only in 1895 that the title of Liskeard & Looe Canal Company was dropped) and the Great Western main line. At the same time the Liskeard & Caradon ceased to work the line, and was itself leased by the Liskeard & Looe. In 1909 the Great Western began to work the railway, and acquired it in 1923.[20]

Though it had been intended to work the canal as well as the railway, the waterway seems in fact to have become disused soon after the latter was opened, except for the broad section from Terras Pill to beyond Sandplace. This section had to be kept navigable under an old agreement with the local landowner, which permitted him toll-free access for craft carrying goods for himself and his tenants. For others, the charge was 3d a boat, and as boatmen carrying seaweed were petitioning against this charge in 1885 on the grounds that the trade could not bear it, margins must have been exiguous. In 1909 about 200 small boatloads (of $1-1\frac{1}{2}$ tons) of sand, seaweed and manure were carried, but later this part also ceased to be used.

Par Canal

Lastly, to a canal built in railway times, and in connection with a railway. J. T. Treffry was a rich man who wished to develop the mineral property he owned in Cornwall, and who planned to join the north coast at Padstow with the south coast by transport that

would tap the many mines on the way. In 1815 he had had surveyed a route from Padstow to Fowey for a possible canal, but
found it impracticable. A good deal later the idea was revived, but
this time using a tramway for most of the route, and a canal only
at the southern end, now placed at Par.

The canal was constructed from the sea harbour there to Pontsmill, where a basin was built at the foot of the Carmears incline of
the Treffry tramway. There were also works at Pontsmill, doubtless
directly served by the canal. It was first used on 4 April 1847,
though the Carmears incline did not come into use till 18 May. Its
length was 1⅞ miles, with a depth of 6 ft at Par and 8 ft at Pontsmill. There was one lock about half-way.

Small containers 6 ft 4 in long, 4 ft 1 in broad and 2 ft 9 in
deep, were used to load tin and lead ores, as well as clay, brought
down in tramway trucks and then transhipped to the canal boats.
As the output of the up-country mines increased, so did the awkwardness of this transhipment. The tramway was therefore extended to Par harbour and opened by August 1855, though there
still seems to have been some use of the canal.[21] In 1873 the Cornwall Minerals Railway, which had taken over the tramway from
the Treffry estate, was building its line from Par to Fowey, and
found it necessary to close the canal. Because its water was needed
for sluicing the harbour and for use by the local lead smelting
works, it was not blocked up at the point where the new railway
crossed it, but carried under the line in two large pipes.

Helston Canal project

In the summer of 1796 Robert Fulton, the American engineer
and inventor, who had for the last three years been corresponding
with Lord Stanhope about the proposed Bude Canal, with a surveyor, Charles Moody, studied the possibilities of a tub-boat canal
nearly 14 miles long across Cornwall from the Helford river near
Gweek to the Hayle river below St Erth. Perhaps he had been
chosen because his book, *A Treatise on the Improvement of Canal
Navigation*, which advocated small canals using inclined planes,
had been published earlier that year. His report on the Helston
Canal was made to a committee under the chairmanship of Sir
Francis Buller, who were interested in a means of carrying copper
from the mines of the interior, and coal to their steam-engines,
and also in conveying sea-sand to improve the local farms.

Fulton planned his canal to leave the Hayle river near St Erth
bridge, mount an inclined plane, and run level to Bosence. There

would be a second plane here, a third at Drym and a fourth at
Nancegollan to lift the waterway to its summit level, which would
extend to Trannack. Here a fifth plane would take it down to run
a little north of Helston to Mellangoose. From a sixth plane there,
it would run to the last incline near Gweek and end in the Helford
river.

The proposed vertical rises of his planes were:

	ft	ft
St Erth	64	
Bosence	105	
Drym	107½	
Nancegollan	73	
		349½
Trannack	142	
Mellangoose	125	
Gweek	82½	
		349½

Fulton estimated his canal at £32,000, to yield a revenue of
£2,268, or 7½ per cent, not counting any sea-to-sea traffic. To begin with, however, he proposed that the canal should first be
built from St Erth to Binner mines near Drym (this was likely to
be the most profitable section, yielding 12 to 15 per cent on cost)
'and there rest till time has exhibited the utility of such works, and
rendered them in some degree familiar to the neighbouring
country'. His argument for the canal, however, was based not
only on its industrial uses, but also on its value in developing
agriculture.[22]

Hayle–Camborne Canal scheme

In 1801 there was a proposal for a canal 7⅝ miles long from
Hayle by way of Angarrack to 'the Manor Mine, then on to Huel
Hope, and finish at a point 28 feet below Carwinin* Bridge, where
there is an ample supply of water, and a canal may be easily conducted to the mines in Camborne, etc.'[23]

Padstow–Lostwithiel scheme, and the Polbrock Canal

Broad, short craft carrying up to 15 tons could with difficulty
get up the River Camel to Polbrock, and up the Fowey to Lostwithiel, on spring tides. The historian, Dr Borlase, had suggested
a possible canal line to connect the two rivers, from Dunmere on

* Carwynnen, 1½ miles south of Camborne.

XV. Inclined planes on the Bude Canal in 1954: (*above*) Marhamchurch; (*below*) Werrington, where the plane crosses a road

XVI. (*above*)   Incline-keeper's house and canal buildings at the head of Hobbacott Down plane in 1965; (*below*) the Liskeard & Looe Union Canal at Landlooe-bridge in 1967. A canal lock, the original road bridge over the lock tail, and a later bridge over canal and railway the Camel to the Fowey near Lanhydrock, down which there would be access to Lostwithiel and Fowey.

Then in September 1793 a public meeting at Bodmin, with Sir William Molesworth in the chair and John and Joseph Edyvean, perhaps sons of old John Edyvean, present, resolved 'that the having an Inland Canal, from the Wadebridge River to the Fowey River, will be of great general advantage to the public, and particularly to the middle part of the country of Cornwall'. A subscription was opened, and Molesworth was asked to get such a scheme surveyed and estimated. George Bentley and Thomas Bolton, whom we have met at work on the Public Devonshire and the Tamar Manure Navigation, were called in. They worked out a course: this could either be short, and depend at either end upon improved river navigations so that boats could move at all times, or could be extended at either end to Wadebridge and to below Lostwithiel. It could also be of different sizes according to the craft seen as using it.[24]

They reported in 1794, suggesting that the river navigations should be improved, and linked by an 8¾-mile-long canal, 20 ft wide at surface, with locks able to take barges 42 ft by 14 or 15 ft, which could also work down the rivers to Padstow and Fowey. Their estimate was £52,767, with £8,069 more for two possible branches. Their figures showed that, relating cost to probable income, the line from the Camel to Bodmin was the more attractive section, at a cost of £21,283, and with an estimated net revenue of £2,654.

Their plans were sent to John Rennie for his comments, which were given in 1796. He favoured a canal using pairs of narrow boats 35 ft long and 7½ ft wide, running right to Wadebridge. But, while he saw a useful traffic in carrying 'the excellent Shell Sand found on the Sea Shore near Padstow' to the waste lands round and to the south of Bodmin, he could see little coast-to-coast traffic through to the Fowey river on a canal so near Land's End.

He therefore suggested one from Guinea Port near Wadebridge to Bodmin, which, if anyone wished, could later be extended to the Fowey river. It would have a river lock, one at Polbrock, three at Boscarne, and a flight near Bodmin. His estimate was £10,498 for Guinea Port to Boscarne, £16,669 for Boscarne to Bodmin, and £3,578 for a feeder, £30,745 in all.

The promoters, headed by Sir William Molesworth, must have agreed in principle, but considered that a good deal of money could be saved if the canal ended at Dunmere, about 1½ miles from Bodmin. Under Rennie's direction James Murray surveyed this. It

M

became the Polbrock Canal project, from near Wadebridge along the western side of the river past Polbrock to Dunmere, with a branch to near Ruthernbridge. The main line was to be 5½ miles long, the branch ¼ mile. The purposes were to supply Bodmin with coal and the area with sea-sand, and to carry mine produce back to be shipped at Wadebridge. An Act[25] authorized this line in 1797, Joseph and John Edyvean being listed as supporters. The capital was to be £18,000 in £50 shares, with £12,000 more if necessary; the authorized tolls were fairly high, 2d per ton per mile for sand and stone, 3d for coal, culm and lime, and 6d for merchandise. Unusually, tolls were also stated for the use of the towpath by cattle, etc. Probably boats of about the size Rennie had recommended, carrying perhaps 15 tons, would have been used, for a minimum charge for 8 tons on a loaded boat was authorized in the Act.

It was a difficult time financially, and Hitchins & Drew, in their *History of Cornwall*, say also that 'when the ground came to be measured with more exactness, that it would be necessary to tunnel through a hill, which would be attended with considerable expense'.[26] Apart from a notice calling a shareholders' meeting after the Act had been passed, nothing more is heard of the project.

In 1825 Marc Isambard Brunel investigated the Padstow–Fowey route for a ship canal. He proposed one 13 miles long, with a large tunnel (always a problem on a canal for seagoing craft) through the high ground near Lanhydrock, to cost £450,000. But it was not thought likely that enough revenue would be earned.[27] Later, after William James had projected a Wadebridge–Bodmin–Fowey railway, the Bodmin & Wadebridge Railway was opened in July 1834 over a very similar route to that of the Polbrock Canal, from Wadebridge to Dunmere and on to Bodmin, with a branch to Ruthernbridge three months later.

The Retyn and East Wheal Rose scheme

From 1819 onwards East Wheal Rose was a prosperous lead and silver mine. In 1821 Silas E. Martin of Crantock promoted a canal from the River Gannel above Newquay to Retyn near St Enoder, with a branch through Trerice to Trewerry mill, Benny and Nanhellan to Ingram Water near the mine, the objects being to supply sea-sand to at least 10,000 acres of land, much of it waste, and carry lead from the mine.

John Edgcumbe did the survey, and estimated the cost at

£16,000 and the yield at over £3,000 pa. But no action followed, and transport to and from the mine was by road until the railway opened in 1849.[28]

AUTHOR'S NOTES & ACKNOWLEDGEMENTS

My thanks are due to very many who have helped me: to Mr David St John Thomas for lending me his file of notes on the Bude Canal; Mr Frank Seekings for reading files of Somerset newspapers; Mr F. C. Dredge, clerk to the Bude-Stratton Urban District Council, for much help; Mr B. A. A. Knight of the Chard History Group; the editor of the *Western Gazette* for permission to consult the files of the *Sherborne Mercury*; Dr A. M. Boyd of Glastonbury, who put material about the Glastonbury Canal, Pillrow Cut and River Axe at my disposal; the chief engineer of the Somerset River Authority and Mr D. R. May; Mr H. L. Douch of the Royal Institution of Cornwall; Mr C. R. Clinker; Mr K. R. Clew; Mr R. A. Atthill; Mr Frank Booker; Mr T. R. Harris; Mr George Ottley; Mr F. Gregson; Mr A. H. Slee; Mr H. G. Kendall; and Mr M. C. Ewans.

I should also like to thank the staffs of British Transport Historical Records, the House of Lords Record Office, the Public Record Office, the Institution of Civil Engineers, County Record Offices, and local libraries and museums for giving me so much help.

My thanks are due to the following for permission to reproduce photographs and other illustrations: B. Chapman, Plate 1(a); A. P. Voce, 1(b); Somerset River Authority, 2(a), 7(a); Conservators of the River Tone, 2(b), Figs 5 and 6; Messrs Douglas Allen, Bridgwater, 3(b); J. V. Smith, 4(a), 5(a), 9(a), 16(a); L. Hoskins, 5(b); D. Milton, 6(a), 6(b); F. C. Perrott, 7(b), 13(a), 16(b); R. J. Sellick, 8(a), 8(b), 11(b), 14(b), 15(b); B. W. Brass, 9(b); H. J. Compton, 10(a), 10(b); Messrs Arthur L. Brinicombe, Exeter, and Exeter Corporation, 11(a); Frank Booker, 12(a); H. G. Kendall, 12(b); House, and the Bude-Stratton Urban District Council, 14(a); Exeter Corporation, jacket picture.

NOTES

Notes to Chapter II

1. For the canal's early history, see E. A. G. Clark, *The Ports of the Exe Estuary, 1660–1860* (1960), and his Ph.D. thesis (University of London), *The Estuarine Ports of the Exe and Teign, with special reference to the Period 1660–1860: a Study in Historical Geography* (1956). See also W. B. Stephens, 'The Exeter Lighter Canal, 1566–1698', *Journal of Transport History* (May 1957); and P. C. de la Garde, 'Memoir of the Canal at Exeter from 1563 to 1724, with a continuation from 1819 to 1830', *Mins. Proc. Inst. Civil Engs.* (1845).
2. Hooker, *The Description of the Citie of Excester*, ed. (1919–47).
3. *Devon Notes and Queries*, vol. 6.
4. James Cossins, *Reminiscences of Exeter Fifty Years Since*, 2nd ed. (1878), p. 54.
5. E. T. MacDermot, rev. C. R. Clinker, *History of the Great Western Railway* (1964), II, p. 127.
6. *Woolmer's Exeter & Plymouth Gazette*, 6 November 1847; *Exeter Flying Post*, 29 June, 6 July 1848.

Notes to Chapter III

1. From the Tone Act of 1699, 10 & 11 Will. III, c. 8.
2. *Journeys of Celia Fiennes*, ed. C. Morris (1947).
3. Daniel Defoe, *A Tour through England and Wales*, Everyman ed., I, p. 269.
4. Quo. E. Jeboult, *A General Account of West Somerset—the Valley of the Tone*, pp. 113 and 117.
5. Conservators' Minute Book, 16 October 1778.
6. Ibid., 16 July 1799.
7. *Sherborne Mercury*, 6 May 1793.
8. Ibid., 7 January 1793.
9. Ibid., 14 January 1793.
10. *Exeter Flying Post*, 14 November 1793.
11. *Felix Farley's Bristol Journal*, 21 December 1793.
12. *Sherborne Mercury*, 23 June 1794.
13. MS. *Reports* of John Rennie (Inst. Civil Engs.).
14. 51 Geo. III, c. 60.
15. *Taunton Courier*, 25 September 1822.
16. *Case of the Conservators of the River Tone and the Inhabitants of Taunton in opposition to a Bill . . .* (1824).
17. 5 Geo. IV, c. 120.

Notes to Chapter IV

1. See deposited plans Nos. 4 and 5, Devon C.R.O.
2. See deposited plans Nos. 6 and 6A, Devon C.R.O.

3. Chard Canal: printed report.
4. Advertisement, 25 March 1824, in an unidentified newspaper.
5. *Grand Ship Canal from Bridgwater to Seaton . . .* (1823).
6. *Gentleman's Magazine* (1824), pt. I, p. 633.
7. Unidentified newspaper report.
8. Records of the Conservators of the River Tone.
9. 6 Geo. IV, c. 199.
10. *Taunton Courier*, 13 July 1825.
11. *Bristol Gazette*, 18 October 1809.
12. *Taunton Courier*, 27 August 1845.
13. Ibid., 8 October 1845.

Notes to Chapter V

1. *Taunton Courier*, 3 January 1827.
2. Conservators' Minute Book, 28 August 1830.
3. Ibid., 3 November 1831.
4. *Taunton Courier*, 12 September 1832.
5. E. F. Goldsworthy, *Recollections of Taunton*, 2nd ed. (1883).
6. E. Jeboult, *A General Account of West Somerset*. Taunton tithe map, 1840, Ralph Ham, surveyor, shows the cut; the O.S. plan of 1886 shows the locks. It is not clear whether the connection with the Grand Western was ever made; such a junction would have required locks or a lift.
7. 7 Will. IV, c. 11.
8. *Sherborne Journal*, 1 April 1841.
9. *Taunton Courier*, 1 October 1845.
10. Bristol & Exeter Railway Minute Book, 3 July 1850.
11. *The Times*, 3 June 1853.
12. Bristol & Exeter Railway Minute Book, 25 August 1853.
13. Conservators' Minute Book, 19 July 1866.

Notes to Chapter VI

1. House of Lords Record Office.
2. 4 Will. IV, c. 53, 16 June 1834.
3. *Sherborne Journal*, 17 September 1840.
4. Ibid., 18 March 1841.
5. *Taunton Courier*, 21 May 1841.
6. *Sherborne Journal*, 5 August 1841.
7. *Chard Union Gazette*, 5 April 1841.
8. *Sherborne Journal*, 10 February 1842.
9. *Woolmer's Exeter & Plymouth Gazette*, 28 May 1842.
10. *Sherborne Journal*, 2 June 1842.
11. Ibid.
12. Ibid., 28 July 1842.
13. Paul N. Wilson, 'Early Water Turbines in the United Kingdom', *Transactions Newcomen Soc.*, XXXI, 1957–9
14. *Sherborne Journal*, 10 August 1843.
15. 9 & 10 Vic., c. 215.
16. 10 & 11 Vic., c. 175.
17. 16 & 17 Vic., c. 192.
18. Inf. from Mr B. A. A. Knight of the Chard Local History Group.
19. Chard Railway Company Minute Book, 14 November 1859.

Notes to Chapter VII

1. I am indebted to Dr A. M. Boyd of Glastonbury for information on the Rivers Axe and Brue.
2. F. A. Knight, *Seaboard of Mendip* (1902), p. 350.
3. E. Galton, 'An Account of Improvement of a Shaking Bog at Meare, Somerset', *Journal of the Royal Agricultural Society* (1845), VI, pp. 182–7.
4. I am indebted to Mr D. R. May, divisional engineer of the Somerset River Authority, for the information on these canals.
5. For material in the early part of this account I am indebted to Dr A. M. Boyd of Glastonbury.
6. *Taunton Courier*, 9 August 1825.
7. Rennie's MS. *Reports* (Inst. Civil Engs.).
8. *Taunton Courier*, 28 February 1827.
9. Other large subscribers from Glastonbury were George Bond (£1,200), Richard Chapman (£1,000), John Vincent (£1,000), John Bulleid (£1,000), and John James Roach (£1,500).
10. 7 & 8 Geo. IV, c. 41.
11. *The Alfred*, Monday, 2 September 1833, quoting the *Dorset County Chronicle*. The 'Monday' of the quotation was therefore almost certainly 19 August.
12. *Bridgwater Times*, 5 December 1850.
13. 11 & 12 Vic., c. 28.
14. *Bridgwater Times*, 21 November, 5 December 1850.
15. 15 Vic., c. 58.
16. The information in this paragraph is taken from D. M. Ross's *Langport and its Church*, 1911.
17. Information given by Mr J. Stevens Cox of Ilchester.
18. *Sherborne Mercury*, 18 February 1793.
19. Ibid., 20 October 1794.
20. 35 Geo. III, c. 105.
21. *Sherborne Mercury*, 29 June, 27 July, 24 August, 21 December 1795; 4 January, 3 October 1796, 23 January 1797.
22. 6 & 7 Will. IV, c. 101.
23. *Report of the Select Committee on the Parrett Navigation Petition*, 1 May 1839.
24. Parrett Navigation Committee Minute Book, 26 June 1840.
25. Ibid., 16 August 1842.
26. Ibid., 22 August 1845.
27. Ibid., 23 August 1848.
28. Ibid., 27 August 1852.
29. Ibid., 17 August 1854.
30. Ibid., 23 August 1855.
31. Ibid., 20 March 1871.
32. Minute Book of the Somersetshire Drainage Commissioners, 22 December 1880.
33. D. M. Ross, *Langport and its Church* (1911), p. 19.
34. *Sherborne Mercury*, 14 January 1793.
35. Ibid., 9 September 1793.
36. Ibid., 1 June 1795.
37. 36 Geo. III, c. 47, 24 March 1796.
38. Pamphlet: *Dorset & Somerset Canal*, undated (c. 1824), B.T.Hist. Recs.
39. Charles Hadfield, *The Canals of the West Midlands* (1966), p. 174.
40. Charles Hadfield, *The Canals of Southern England* (1955), pp. 152–4.
41. Patent No. 2284 (1798).
42. *Bath Chronicle*, 16 October 1800.
43. *Sherborne Mercury*, 15 March, 5 April 1802.
44. *Exeter Flying Post*, 5 January, 2 February 1826.
45. For a full account of the canal's remaining works, see Robin Atthill's chapter 'The Dorset & Somerset Canal' in his *Old Mendip* (1964).

Notes to Chapter VIII

1. *Exeter Flying Post*, 4 October 1792.
2. See my *The Canals of South Wales and the Border*, 2nd ed. (1967), pp. 132–5.
3. The principal documents covering this earlier period are in the Muniment Room of Exeter corporation.
4. 36 Geo. III, c. 46.
5. *An Authentic Description of the Kennet & Avon Canal* (1811).
6. W. Harding, *Historical Memoirs of Tiverton* (1845–7).
7. Counsel's brief of 1864, when the canal was sold to the G.W.R.
8. Summons to Meeting, 18 March 1830.
9. These were: Combe Hay on the Somersetshire Coal Canal (see my *The Canals of Southern England* (1955), pp. 152–4); Ruabon on the Ellesmere Canal (see my *The Canals of the West Midlands* (1966), p. 174); Mells on the Dorset & Somerset Canal (see p. 93 of this book); Tardebigge on the Worcester & Birmingham Canal (see my *The Canals of the West Midlands* (1966), pp. 140–2); and Camden Town on the Regent's Canal (see my *The Canals of the East Midlands* (1966), pp. 129–30).
10. *Transactions Inst. Civil Engs.* (1838), II, p. 185.
11. *Woolmer's Exeter & Plymouth Gazette*, 16 June 1831.
12. *Taunton Courier*, 25 February 1835.
13. Ibid., 4 July 1838.
14. Bristol & Exeter Railway Minute Book, 5 December 1845.
15. *Western Times*, 4 September, 27 November 1847.
16. Bristol & Exeter Railway Minute Book, 22 March, 5 April 1848; *Exeter Flying Post*, 8 June 1848; *Woolmer's Exeter & Plymouth Gazette*, 24 June, 15 July 1848.
17. Bristol & Exeter Railway Minute Book, 9 July 1856.
18. 27 & 28 Vic., c. 184.

Notes to Chapter IX

1. *A Report by the Committee of the intended Public Devonshire Canal at a meeting at Crediton, 17 October 1793*, Devon C.R.O., and printed plan, Exeter corporation, Muniment Room.
2. *Exeter Flying Post*, 17 June 1824.
3. Moore, *History of Devonshire*, I, p. 438.
4. This account is partly based on M. C. Ewans, *The Haytor Granite Tramway and Stover Canal* (1966), and E. A. G. Clark's MS. Ph.D. thesis, *The estuarine ports of the Exe and the Teign* (University of London, 1954).
5. 32 Geo. III, c. 103.
6. *Woolmer's Exeter & Plymouth Gazette*, 18–23 November 1829.
7. Ibid., 22 July 1848.
8. Deposited plan, Devon C.R.O.; *Sherborne Mercury*, 28 January, 25 February, 23 September 1793; 13, 27 January, 17 February 1794.
9. John Smeaton, *Reports*, II, 240–1.
10. Plymouth & Dartmoor Railway Proprietors' Minute Book, 3 January 1826.
11. Ibid., 8 December 1828.
12. Ibid., 3 January 1829.
13. Vol. I, p. 73.
14. H. P. R. Finberg, *Tavistock Abbey* (1951), pp. 44–5.
15. Plans, printed estimate, and report, Devon C.R.O.
16. *Sherborne Mercury*, 10 August 1795.
17. 36 Geo. III, c. 67, 26 April 1796.
18. *Sherborne Mercury*, 27 June 1796.

19. Ibid., 15 August 1796.
20. John Rennie's MS. *Reports* (Inst. Civil Engs.).
21. *Sherborne Mercury*, 12 January 1801.
22. Ibid., 15 February 1802; *Exeter Flying Post*, 9 May 1811.
23. *Tavistock Canal*, report of the meeting on 16 March 1803.
24. 43 Geo. III, c. 130.
25. *Exeter Flying Post*, 4 April 1811.
26. Tonnage figures from J. H. Collins, *Observations on the West of England Mining Region* (1912).
27. R. Hansford Worth, 'Early Western Railroads', *Trans. Plymouth Inst.*, vol. 10, 1887–90.
28. C. von Oeynhausen and H. von Dechen, 'Report on Railways in England in 1826–27', *Trans. Newcomen Soc.*, XXIX, 1953.
29. Moses Bawden, 'Mines and Mining in the Tavistock District', *Trans. Devonshire Association*, vol. 46, 1914; and W. White, *History, Gazetteer and Directory of the County of Devon* (1850).
30. Tavistock Canal, annual report, 27 September 1816.
31. *Plymouth & Dock Telegraph*, 28 June 1817.
32. Rachel Evans, *Home Scenes, or Tavistock and its Vicinity* (1846).
33. Tavistock Canal report, 19 August 1859.
34. Ibid., 14 August 1860.
35. Tavistock Canal Minute Book, 10 September 1866.
36. Ibid., 17 May 1870.
37. For a full account of the Tamar, the Tavistock Canal, and the neighbouring mines, see Frank Booker, *The Industrial Archaeology of the Tamar Valley* (1967).
38. G. M. Doe, *Old Torrington Landmarks* (1931), p. 12.
39. *Exeter Flying Post*, 5 December 1793.
40. MS. letter to the clerk of the Salisbury & Southampton Canal.
41. Devon C.R.O., deposited plan.
42. *Exeter Flying Post*, 7 March 1823.
43. Ibid., 23 December 1823, 18 March, 26 August 1824.
44. Ibid., 19 August 1824.
45. *Exeter Flying Post*, 15 February 1827.
46. Inkerman Rogers, *A Record of Wooden Sailing Ships . . . built at the Port of Bideford from 1568 to 1938* (1947).
47. Article, 'The Bideford and Okehampton Railway of 1831', in *Trans. Devonshire Association*, vol. 34, 1902.
48. 'North Devon Fifty Years Ago', *North Devon Journal*, September 1906.
49. *North Devon Herald*, 21 July 1870.
50. Ibid., 13 June 1872.
51. See deposited plans, Devon C.R.O. and *Exeter Flying Post*, 6 September 1810, 8 July 1813.
52. I am indebted to Mr A. H. Slee for information on the Braunton canalization.
53. See plan, Exeter corporation Muniment Room.

Notes to Chapter X

1. *Sherborne Mercury*, 14 March 1774.
2. 14 Geo. III, c. 53, 24 May 1774.
3. *Sherborne Mercury*, 5 December 1774.
4. Rees's *Cyclopaedia*, article 'Canals' (1805).
5. For the information about Robert Fulton I am indebted to H. W. Dickinson's *Robert Fulton, Engineer and Artist* (1913), chs. III and IV.
6. *Report of the Committee appointed to conduct a Survey* (1818).
7. *Exeter Flying Post*, 5 August 1819.
8. Ibid., 17 July 1823.
9. Ibid., 13 May 1824.

10. *Report of George Call, Chairman to the Committee, before the General Annual Assembly on 2 May 1827.*
11. H. P. R. Finberg, *Tavistock Abbey* (1951), pp. 89–90.
12. MS. *Report* to the Grand Western Canal Committee upon the Lifts and Inclined Plane at Wellisford.
13. *Report* to the Exchequer Bill Loan Commissioners, 31 May 1838.
14. Monthly *Reports* of Chairman to the Committee, from which the two following quotations are also taken.
15. Resident engineer's *Report* to chairman, 20 November 1830.
16. Ibid., 23 November 1830.
17. Ibid., 16 December 1830.
18. A. F. Robbins, *Launceston Past and Present* (1888), gives further information on this project.
19. In the possession of the Rev. Arthur J. Sleeman.
20. 54 & 55 Vic., c. 75.
21. 1 Edw. VII, c. 258.
22. 8 & 9 Eliz. II, c. 28.

Notes to Chapter XI

1. H. M. Stocker's contribution to the 1852 *Report* of the Penzance Natural History & Antiquarian Society, p. 89, records the 'recent' discovery of the boats. He writes, 'they have not been seen for at least a hundred and twenty years'. Hitchins & Drew, in their *History of Cornwall* (1824), speak of the canal being used 'for several years'. From these statements I have provided the approximate datings given in the text.
2. 13 Geo. III, c. 93, 1 April 1773.
3. Quoted from the Act.
4. J.H.L., 12 March 1773.
5. This account owes much to information supplied by Mr G. White of Newquay, and to researches made by the Rev. W. Jago in 1914.
6. *Sherborne Mercury*, 26 July, 23 August 1773; 2 January, 12 June, 10 July 1775; 26 August, 2 September 1776; 11 August 1777; 25 January 1779.
7. *The Monthly Review* (1791), pp. 119–20.
8. A. Rees, *Cyclopaedia*, article 'Canals', 1805.
9. *Sherborne Mercury*, 25 May 1795.
10. *The Monthly Review*, loc. cit.
11. *Exeter Flying Post*, 21 August 1823.
12. *Report of James Green, Civil Engineer, on a Canal, Rail Road, and Turnpike Road, between the Ports of East and West Looe, and the Town of Liskeard, in the County of Cornwall* (Exeter, 1823).
13. *Sherborne Mercury*, 11 December 1797.
14. Liskeard & Looe Union Canal Minute Book, 19 July 1825.
15. Ibid., 2 September 1825. Sick clubs for canal navvies were often organized, but the retention of a doctor is unusual.
16. Cornwall Record Office, Buller collection, No. 407.
17. Liskeard & Looe Un'on Canal Minute Book, 2 February 1836.
18. Ibid., 23 March 1841.
19. Ibid., 26 December 1845.
20. For an account of both railways, see D. S. Barrie & 'Precursor', From Looe to the Cheesewring', *Railway Magazine*, December 1935; and also J. M. Tolson, 'Along the East Looe Valley', *Railway Magazine*, May 1967.
21. I am indebted to Mr. C. R. Clinker for most of my information on this canal.
22. Robert Fulton, *Report on the Proposed Canal between the Rivers Heyl and Helford*, 24 November 1796.
23. Information from Mr T. R. Harris of Camborne, quoting descriptive notes to a plan drawn by Charles Moody in 1801.

24. Cornwall C.R.O., deposit D.D.X. 266.
25. 37 Geo. III, c. 29.
26. Most of the material for this account is taken from John Rennie's MS. *Reports* (Inst. Civil Engs.).
27. R. Beamish, *Memoir of the Life of Sir Marc Isambard Brunel* (1862).
28. H. L. Douch, *East Wheal Rose.*

APPENDIX I

Summary of Facts about South Western Canals and Navigations

A. *Rivers Successfully Made Navigable*

River	Date of Act under which Work was begun	Date Wholly Opened	Approx. Cost at Opening (£)	Terminal Points
Parrett (and Westport Canal)	1836	1840	39,000	Burrow Bridge-Westport br. Thorney
Tone	1699	1717	5,700	Ham Mills–Taunton

[1] 10½ miles river including Thorney branch, 2 miles Westport Canal. The River Ivel (Yeo) was also regularly used for 3¾ miles to Load bridge, though not part of the navigation. (See Ivelchester & Langport Navigation.)
[2] The largest boats seems to have carried about 24 tons. Size of locks 73 ft 6 in × 16 ft 0 in approx.

B. *Rivers with Uncompleted Navigation Works*

River	Date of Act under which Work was begun	Money Spent (£)	Terminal Points Authorized
Ivelchester & Langport[1]	1795	6,000	Ilchester–Langport
Tamar Manure[4]	1796	11,000	Morwellham–Tamerton Bridge

[1] Part canalized drain, part canal, part river.
[2] Probably little, if any, work was done on the river section.

Length	Greatest Number of Locks	Size of Boats Taken	Date of Disuse for Commercial Traffic	Date of Abandonment	Whether bought by Railway, and Present Ownership
12½ miles[1]	4 1 half-lock	NK[2]	c. 1878	1878	No. transferred to Somersetshire Drainage Commissioners, 1878
3 miles[3]	4 locks 4 half-locks	54 ft × 13 ft	c. 1878[4]	1967	Bought by the Bridgwater & Taunton Canal in 1832 for about £10,000. Owned by B.W.B.

[3] The length of the Tone from Firepool, Taunton, to Burrow Bridge was 11⅝ miles, or 17⅝ miles to Bridgwater, but the navigation works were on the uppermost 3 miles. From Firepool the river was also navigable for ¾ miles to near French Weir, whence for a short time a ½-mile cut with two locks ran to the Grand Western Canal.
[4] Except for Firepool upwards for about ¾ miles, 1907, and Ham–Burrow bridge, 1929.

Length on which Work was done	Greatest Number of Locks	Size of Boats Taken	Date of Disuse for Commercial Traffic	Date of Abandonment	Later Events
8¼ miles[2]	7	75 ft × 7½ ft[3]		c. 1797	River Ivel (Yeo) continued to be used in unimproved state.
2¾ miles Morwellham to Newbridge (Gunnislake)	1	70 ft × 20 ft[3]	Closed c. 1929		

[3] Lock size.
[4] Part river, part canal.

C. *Canals, the Main Lines of which were Completed as Authorized*

Canal	Date of Act under which Work was begun	Date wholly Opened	Approx. Cost at Opening £	Terminal Points	Branches Built
Bridgwater & Taunton	(1811) 1824 1837[1]	1827 1841[2]	97,000 plus 100,000[3]	Taunton–Huntworth (later Bridgwater Dock)	—
Brown's	1801	NK	NK	R. Brue–North Drain	—
Cann Quarry	None	1829	NK	Cann Quarry–Marsh Mills	—
Chard	1834	1842	*c.* 140,000	Creech St Michael–Chard	—
Exeter	1539	1566 1701, 1830[7]	5,000 NK[8] 113,000	Turf[9]–Exeter	—
Galton's	1801	NK	NK	River Brue–North Drain	—
Glastonbury	1827	1833	*c.* 30,000	Highbridge–Glastonbury	—
Hackney	None	1843	NK	River Teign–Kingsteignton Road	—

[1] 1811, original Act for Bristol & Taunton Canal; 1824, Act shortened line, but repeated much of 1811 Act; 1837, Act authorized extension from Huntworth and building of Bridgwater Dock.

[2] 1827, Taunton–Huntworth; 1841, extension and dock.

[3] £97,000, cost of original line; £100,000, extension and dock.

[4] 13½ miles Huntworth–Taunton; 1 mile extension to Bridgwater dock.

[5] Including the tide lock between Bridgwater dock and the Parrett, and Firepool lock between the canal and the Tone at Taunton.

[6] The caissons of the inclined planes took one boat at a time; the locks held two.

[7] 1566, opening of original canal; 1701 and 1830, opening of major reconstructions.

Length	Greatest Number of Locks	Size of Boats Taken	Date of Disuse for Commercial Traffic	Date of Abandonment	Whether bought by Railway and Present Ownership
15¼ miles[4]	7[5]	54 ft × 13 ft	c. 1907	—	Bought by railway in 1866 for £64,000. Now owned by British Waterways Board.
1 mile	None	Small boats	NK		
2 miles	None	Small tub-boats	c. 1835	—	Afterwards used as a mill-leat
13½ miles	4 inclined planes. 1 lock 1 stop lock	26 ft × 6 ft 6 in[6]	1868	1867	Bought by railway in 1867 for £5,945 and closed
5 miles[10]	1 on canal, 1 tide lock, 1 side lock[11]	120 ft × 26 ft	Open	—	No. Owned by Exeter Corporation
1⅜ miles	1	Small boats	c. 1850	1897	
14⅛ miles	1 on canal 1 tide-lock	64 ft × 18ft 6in[12]	1854[13]	1852[13]	Bought by railway in 1848 for £7,000. Kept open. Resold to another railway in 1852 for £8,000. Railway laid along bank and in bed at one point
⅝ mile	1 tide-lock	54 ft × 14 ft[14]	1928	—	No. Clifford Estate, Kingsteignton

[8] I do not know the cost of the 1701 improvements. The figure of £250,000 sometimes given is not credible.

[9] After 1827.

[10] Length of canal, 1¾ miles. After 1701 extension, 2¼ miles. After 1830 extension, 5 miles.

[11] See text for changes in locks.

[12] Size of lock.

[13] Except for the lower portion in Highbridge.

[14] The lock would take two boats at once.

C. *Canals, the Main Lines of which were Completed as Authorized*

Canal	Date of Act under which Work was begun	Date wholly Opened	Approx. Cost at Opening £	Terminal Points	Branches Built
Liskeard & Looe Union	1825	1828	17,200	Terras Pill–Moorswater	—
Par	None	1847	NK	Par Harbour–Pontsmill	—
Stover	None	1792	NK	River Teign–Teigngrace	
Tavistock	1803	1817 1819 (branch)	62,000 plus 8,000 for branch	Morwellham–Tavistock	Lumburn aqueduct–Millhill
Torrington (Rolle)	None	1827	c. 40,000	R. Torridge–Torrington	—

[1] Size of locks, other than the entrance lock at Terras Pill, 57 ft × 13 ft 6 in approx.
[2] Except for small section Terras Pill–Sandplace disused about 1910.
[3] Including a staircase pair.

Length	Greatest Number of Locks	Size of Boats Taken	Date of Disuse for Commercial Traffic	Date of Abandonment	Whether bought by Railway and Present Ownership
5⅞ miles	25	50 ft × 10 ft[1]	c. 1861 closed[2]	—	Canal company built railway alongside canal in 1860
1⅞ miles	1	NK	c. 1855	—	Built as part of railway
1⅞ miles	5[3] (incl. river lock)	54 ft × 14 ft	Top ⅝ mile c. 1867 whole c. 1939	—	Bought by railway in 1862 for £8,000
Main line 4 miles Branch 2 miles	2 inclined planes	NK, small tub-boats	c. 1832, Millhill branch c. 1873, main line[4]	—	No. Bedford Estate, Tavistock
6 miles	1 tide-lock 1 inclined plane	NK (tub-boat)	c. 1871	—	Part of site sold to railway

[4] Most of the old canal is used for generating hydro-electric power.

D. *Canals, the Main Lines of which were not Completed*

Canal	Date of Act under which Work was Begun	Date Opened	Approx. Cost at Opening £	Authorized Terminal Points	Terminal Points as Built	Branches Built
Bude	1819	1823[1]	118,000	Bude–Thornbury	Bude–Blagdonmoor	Red Post–Druxton; navigable feeder to Virworthy
Grand Western	1796	1814 1838[4]	244,500 +c. 105,000[5]	Taunton–Topsham	Taunton–Burlescombe	Tiverton[6]
St. Columb	1773	c. 1777–9	NK	Trenance Point–Lower St. Columb Porth	See text	—

[1] Except the Druxton branch, opened 1825.
[2] This was the size of the tub-boats. Larger craft worked on the short length of broad canal between Bude and Marhamchurch, the locks of which were 63 ft × 14 ft 7 in.
[3] Broad canal section.
[4] Tiverton–Lowdwells 1814. Lowdwells–Taunton 1838.

E. *Canals Partly Built but not Opened*

Dorset & Somerset Canal	Authorized 1796, from Kennet & Avon Canal at Widbrook (near Bradford-on-Avon) to Gains Cross near Sturminster Newton, with branch to Nettlebridge. Nearly 8 miles of branch built up to 1802, and about £66,000 spent.
Exeter & Crediton Navigation	Authorized 1801, from Exeter to Crediton. About ½ mile cut near Exeter, 1810–11. Cost not known.

Length	Greatest Number of Locks	Size of Boats Taken	Date of Disuse for Commercial Traffic	Date of Abandonment	Whether bought by Railway and Present Ownership
35½ miles	2 and sea lock 6 inclined planes	20 ft × 5 ft 6 in²	1891 tub-boat canal, c. 1900 barge canal. Closed except for basin	1891 1960³	Transferred to Stratton & Bude U.D.C. 1901. Now North Devon Water Board and Bude–Stratton U.D.C.
11 miles Tiverton–Lowdwells; 13½ miles Lowdwells–Taunton	2⁷ 7 lifts 1 inclined plane	26 ft × 6 ft 6 in.	1867, Lowdwells–Taunton. Remainder c. 1924	1864 Lowdwells–Taunton 1962 Lowdwells Tiverton	Leased to railway 1854. Sold to railway for £30,000 1864. Now owned by British Waterways Board
c. 6½ miles	Probably 2 inclined plane	NK (tubboats)	c. 1781	—	No

⁵ Tiverton–Lowdwells, £244,500 (barge canal); Lowdwells–Taunton c. £105,000 (tubboat canal).
⁶ Only a short part of the main line of the barge canal was built; the rest was the Tiverton branch.
⁷ At Lowdwells and Taunton, the latter being a stop lock. I have not counted the locks that were part of the lifts.

F. Canals Authorized but not Begun

APPENDIX II
Principal Engineering Works
A. *Inclined Planes*

Canal	Name of Plane	Vertical Rise	Dates Working	Notes
Tavistock	Morwellham	237 ft	1817– c. 1873	Double-track. Goods transhipped from boats to plane and back. Water-wheel
do.	Millhill	19½ ft	1819– c. 1832	Double-track. Boats carried in cradles. Counterbalanced, with horses as extra motive power
Bude	Marhamchurch	120 ft	1823– 1891	Double-track. Boats had wheels to run on and off the plane. Water-wheel
do.	Hobbacott Down	225 ft	do.	As above, but bucket-in-a-well system, with steam-engine in reserve
do.	Venn (Veala)	58 ft	do.	As above, but water-wheel
do.	Merrifield	60 ft	1825– 1891	do.
do.	Tamerton	59 ft	do.	do.
do.	Werrington	51 ft	do.	do.
Torrington	Weare Giffard	c. 60 ft	1827– c. 1871	Probably double-track. Probably boats with wheels as on Bude Canal. Water-wheel
Grand Western	Wellisford	81 ft	1838– 1867	Double-track. Boats carried in cradles. Designed for bucket-in-a-well system, but steam-engine substituted before opening.
Chard	Thornfalcon	28 ft	1841– 1868	Double-track. Boats carried in caissons. Counterbalanced, with extra weight of water as motive power
do.	Wrantage	27½ ft	do.	do.
do.	Ilminster	82½ ft	1842– 1868	do.
do.	Chard Common	86 ft	do.	Single-track. Boat carried in cradle. Water turbine

B. *Lifts*

Canal	Name of Plane	Vertical Rise	Dates Working	Notes
Dorset & Somerset	Mells	20 ft	1800– 1802	Boats carried floating in two caissons suspended from carrying wheels. Motive power, the adding of water to the descending caisson
Grand Western	Taunton	23½ ft	1838–67	do.
do.	Norton	12½ ft	do.	do.
do.	Allerford	19 ft	do.	do.
do.	Trefusis	38½ ft	do.	do.
do.	Nynehead	24 ft	do.	do.
do.	Winsbeer	18 ft	do.	do.
do.	Greenham	42 ft	do.	do.

C. *Tunnels over 500 yards*

| Tavistock Canal | Morwelldown | 2,540 yd |
| Chard Canal | Crimson Hill | 1,800 yd approx. |

D. *Outstanding Aqueducts*

Bude Canal	Tamar
Chard Canal	Creech (Tone)
Dorset & Somerset Canal	Coleford
Grand Western Canal	Halberton (over railway)
Tavistock Canal	Lumburn
Torrington Canal	Beam (Torridge)

INDEX

The principal references to canals and river navigations are indicated in bold type